Concrete and Clay

Urban and Industrial Environments

Series editor: Robert Gottlieb, Henry R. Luce Professor of Urban and Environmental Policy, Occidental College

Maureen Smith, *The U.S. Paper Industry and Sustainable Production: An Argument for Restructuring*

Keith Pezzoli, *Human Settlements and Planning for Ecological Sustainability: The Case of Mexico City*

Sarah Hammond Creighton, *Greening the Ivory Tower: Improving the Environmental Track Record of Universities, Colleges, and Other Institutions*

Jan Mazurek, *Making Microchips: Policy, Globalization, and Economic Restructuring in the Semiconductor Industry*

William A. Shutkin, *The Land That Could Be: Environmentalism and Democracy in the Twenty-First Century*

Richard Hofrichter, ed., *Reclaiming the Environmental Debate: The Politics of Health in a Toxic Culture*

Robert Gottlieb, *Environmentalism Unbound: Exploring New Pathways for Change*

Ken Geiser, *Materials Matter: Toward a Sustainable Materials Policy*

Matthew Gandy, *Concrete and Clay: Reworking Nature in New York City*

Concrete and Clay

Reworking Nature in New York City

Matthew Gandy

The MIT Press
Cambridge, Massachusetts
London, England

This book was set in Bembo by Achorn Graphic Services, Inc.

Printed and bound in the United States of America.

Library of Congress Cataloging-in-Publication Data

Gandy, Matthew.
Concrete and clay : reworking nature in New York City / Matthew Gandy.
p. cm. — (Urban and industrial environments)
Includes bibliographical references and index.
ISBN 0-262-07224-6 (hc. : alk. paper)
1. Urban ecology—New York (State)—New York. 2. Human ecology—New York (State)—New York. 3. City planning—Environmental aspects—New York (State)—New York. 4. Land use, Urban—New York (State)—New York. I. Title. II. Series.

HT243.U62 N74 2002
304.2′09747′1—dc21

2001054604

Contents

Preface

In October 2001 I had an opportunity to present material from this book as part of a festival of New York history organized by the recently founded Gotham Center in midtown Manhattan. The festival included a documentary film about the city that concluded with the topping-out ceremony for the World Trade Center in 1971. The packed auditorium sat in disbelief as the camera panned across construction workers waving against the backdrop of the Manhattan skyline. Though the pall of acrid smoke from the destruction of the World Trade Center had now receded, the city remained in a deep state of shock. I was struck, however, by the sense of common purpose and civic resilience which bodes well for a city that has coped with immense challenges in the past. Since my first encounter with New York City in the spring of 1990 I have become gradually immersed in a project to try to understand better how a modern metropolis is able to function, exploring a series of relationships between nature, cities, and social power. The experience has forced me to rethink the boundaries between different bodies of urban knowledge, ranging from the voices of grassroots activists to the more abstract domain of engineers, planners, and urban theorists. My fascination with New York stems in part from its contradictory character: most notably the juxtaposition of a sophisticated public sphere with the raw energies of a global financial center. Debates since September 2001 about the reconstruction of lower Manhattan perfectly illustrate this tension: Who, for example, has the legitimacy to speak on behalf of the city? How can public and private interests be reconciled in the face of powerful pressures to ensure that land values are protected and enhanced? And what kind of architectural forms might give expression to collective memory in the context of competing demands for the use of urban space? My writing on New York City has coalesced around the theme of urban nature as the organic web that links the abstract and concrete domains of the city. I argue that the very idea of "modernity" and the modern city is closely tied up with the changing use, meaning, and understanding of nature.

At an early stage in the research I decided to focus my analysis on five specific aspects of New York City: the building of a modern water supply system; the creation and meaning of public space; the design and construction of landscaped roads; the grassroots environmentalism of the urban *barrio;* and the contemporary politics of pollution. Of necessity, some critical themes such as energy, biotic diversity, and regional agricultural change have been left to one side. In writing this book I have been stepping between different worlds using a variety of different data sources. A central element of my research strategy has been the use of oral history in order to capture fragments of institutional memory as a means to better understand the profound changes that have taken place over the last thirty years. My aim is to create a living history of the urban environment as an unfolding and unresolved dynamic. In order to make sense of contemporary developments, it has also been necessary to delve further into the past to reconstruct the context for the creation of new forms of urban nature since the early nineteenth century.

My work could not have been completed without the invaluable contribution of staff at a range of libraries and archives, including the Avery Architecture & Fine Arts Library at Columbia University, the New York Municipal Archives, the Westchester County Archives, the New-York Historical Society, the Museo del Barrio, the Schomburg Center for Research into Black Culture, the Museum of the City of New York, the Centro de Estudios Puertorriqueños at the City University of New York, the New York Public Library, and the Special Archive of the Triborough Bridge and Tunnel Authority. Further sources were provided in London by the British Library, the Institution of Civil Engineers, and the Royal Institute of British Architects. A project of this kind invariably owes an intellectual debt to others. I would like to thank Janet Abbate, Nancy Anderson, Karen Bakker, Mark Bassin, Maribel Biasco, Steven Corey, Richard Dennis, David Golub, Michael K. Heiman, Sharon Kinsella, Matthew Lockwood, Hugh Prince, Laura Pulido, and Virginia Sánchez Korrol for their detailed responses to earlier versions of my manuscript. I must make special mention of Simon Gruber's patience in explaining to me the intricacies of the city's watershed politics, and the unique ambience of the Hungarian Pastry Shop on Amsterdam Avenue for providing such a relaxed environment to mull over ideas. Thanks also to Clay Morgan, Matthew Abbate, and Chryseis Fox at the MIT Press for their patience, diligence, and professionalism at every stage of the project. Cartographic assistance for the illustrations was provided by Elanor McBay, Catherine Dalton, and Susan Rowland. Journal reviewers for *Antipode* and *Transactions* have also played a useful role in sharpening my arguments and analysis for earlier material derived from chapters 1 and 4. Financial support for the research was provided by the Graham Foundation for Advanced Studies in the Fine Arts, the Economic and Social Research Council, and the British Academy. Finally, a special thank you to Todd Alden, Reetu Arora, Billie Chan, Maria Gandy, Mike Gandy, Ian Mansfield, Kakoli Ray, and Alex Veness for all their support and encouragement.

Matthew Gandy
London and New York
November 2001

Concrete and Clay

Introduction

> The chief function of the city is to convert power into form, energy into culture, dead matter into the living symbols of art, biological reproduction into social creativity.
>
> —*Lewis Mumford*[1]

> Crammed on the narrow island the millionwindowed buildings will jut glittering, pyramid on pyramid like the white cloudhead above a thunderstorm.
>
> —*John Dos Passos*[2]

In 1524 the Italian explorer Giovanni da Verrazzano sailed northward along the eastern seaboard of what is now the United States. Verrazzano described a coastline carpeted with "immense forests of trees . . . too various in colours, and too delightful in appearance to be described." As he entered the mouth of the Hudson River and approached Manhattan Island, he found the land to be "much populated," the people "clothed with the feathers of birds of various colours."[3] Though these original landscapes and cultures have been largely obliterated, the Manhattan skyline has often been described in terms of metaphors drawn from nature, the steel and concrete mass of the modern city simply a luxuriant outgrowth of these earlier forests. The architectural ebullience of the modern city has intensified a perception that the scale and dynamism of New York owes more to the raw power of nature than to the prosaic efforts of human labor.

This book explores the different ways in which the raw materials of nature have been reworked in New York City to produce a metropolitan nature quite distinct from the premodern forms encountered by early settlers. The expression "metropolitan nature" captures something of the multiple meanings of modern nature, ranging from the preservation of wilderness for the consumption of an idealized natural beauty to the construction of complex networks for the provision of water. New York City has some of the most important examples of urban nature in North America: its verdant parks and gardens; its landscaped parkways like green rivulets cutting free from the city; and its magnificent water technologies, which have harnessed a regional hydrological cycle to serve the needs of nine million people. It is paradoxically in the most urban of settings that one becomes powerfully aware of the enduring beauty and utility of nature. It is the reshaping of nature that has made civilized urban life possible. Nature has a social and cultural history that has enriched countless dimensions of the urban experience. The design, use, and meaning of urban space involves the transformation of nature into a new synthesis.

The production of urban nature is a microcosm of wider tensions in urban society. In the early decades of the nineteenth century, the rapidly growing settlement of New York City faced the prospect of social and economic collapse. Only a new kind of mediation between nature and the city could avert this looming catastrophe. The construction of the city's water supply system instituted new patterns of municipal intervention and innovative mechanisms for the raising of capital. Later developments such as the creation of Central Park and the construction of landscaped roads all flowed from this emerging urban dynamic. These developments involved a realignment between municipal government, capital, and nature, which set in place a remarkably resilient framework for the construction of urban infrastructure lasting well into the twentieth century. Where technical and political opinion concurred, significant changes in the urban environment could ensue. Yet the public health concerns of nineteenth-century progressives evolved in an uneasy symbiosis with wider demands to make urban space more efficient for the purposes of capital accumulation. A cursory glance at mortality and morbidity statistics for nineteenth-century New York shows that squalid and insanitary conditions for the city's poor persisted long after improvements had been made in urban infrastructure.

In nineteenth-century Europe and North America, the dominance of metropolitan cultural and political power played a decisive role not only in driving the process of urbanization but also in changing perceptions of relations between nature and civil society. The power of nature as an ideological construct not only helped to reinforce perceptions of the city itself but also intensified new forms of social stratification within it.[4] In nineteenth-century New York, for example, competing "cultures of taste" in relation to the reconstruction of cities were coded forms of social and political conflict that revealed intense anxieties about the pace and direction of urban change. Cultural differences were from the outset driven not only by social class and gender but by a series of ethnic tensions and racialized stereotypes that infused the discourses of urban planning and design. The creation of Manhattan's Central Park, for example, presented an Anglophile vision of the English picturesque that was anathema to much Irish political and intellectual opinion at the time. And romantic influences on landscape design fostered suspicions that southern Europeans and other recent immigrants lacked the aesthetic sensibilities to appreciate beauty in nature. Equally, much of the public health discourse of the modern era has drawn on racialized and class-based caricatures of cleanliness and hygiene, stressing behavioral admonitions to the poor in place of any real challenge to the patterns of property ownership and official neglect that underpin the persistence of slum housing. As recently as 1971 an installation by the German artist Hans Haacke that simply chronicled the changing landlords and real estate values for hundreds of dilapidated tenements on Manhattan's Lower East Side was deemed too "controversial" to be shown at the Guggenheim Museum.[5]

The transformation of the experience and perception of nature in New York City intersects with a series of social, political, and economic developments.

The modern experience of nature in the city ranges from new water technologies within the home to the aesthetic discourses of European landscape design. Its history is intimately related not only to upstate landscapes stretching far beyond the outer boundaries of the city as a physical entity, but also to more distant places bound up with the colonial and imperial legacies of the United States. The ecological and political hinterland of New York City has developed into a global arena of power binding the history of the city to ever-widening flows of people, commodities, and ideas. Our understanding of the production of urban nature involves an engagement with a diverse array of discourses ranging from the technological utopias of twentieth-century modernism to the radical political legacies of the late 1960s and early 1970s. Civil rights-inspired social and political struggles might seem to have little to do with urban environmental history, but these developments are indispensable to any analysis that moves us beyond the technical exigencies of city management to a deeper engagement with the social production of urban space. An exploration of the meaning of urban nature demands that the juxtapositions and contradictions of the urban environment be moved to the center stage of analysis. It is a narrative strategy that seeks to challenge existing conceptions of the relationship between nature and the city and bring together insights and ideas from a range of disciplines and perspectives. To bring questions of environmental justice into the frame of analysis compels us to see urban environmental change not simply as a function of technological change or of the dynamics of economic growth but as an outcome of often sharply different sets of political and economic interests. The shared experience of metropolitan nature reveals many intersecting facets between the public and private dimensions of urban space. The cultural hybridity of urban nature warns us against transcendent views of nature as something beyond historical process or exceptionalist views of nature that search for discrete sets of national or cultural characteristics.

Popular ecological accounts of the urban process have tended to view the productivist dynamic of modern societies as necessarily antithetical to environmental well-being. It is assumed that the abandonment of modernity as both a utopian project and a distinctive set of scientific methodologies and philosophical assumptions holds out the hope for a new kind of interaction with nature. The argument presented here moves in a rather different direction. It is suggested that the production of modern cities has altered the relationship between nature and society in a series of material and symbolic dimensions. It is only by radically reworking the relationship between nature and culture that we can produce more progressive forms of urban society. The modification of nature is itself part of the pretext for a more civilized kind of urbanism through which the benefits of metropolitan nature can be spread more widely. The reworking of modern nature is a collective project that applies the human imagination to the transformation of urban space and affirms the interdependencies that sustain a flourishing civic realm. Few environmentalist accounts of the modern city conceive of the urbanization of nature as a historical and political process and fewer still make any distinction between modernity and capitalist urbanization.

There are, of course, alternative modernities that vie for representation in the urban landscape, as different conceptions of meaning and identity are etched into the fabric of the city. We shall see how the struggle to impose particular interpretations of the urban question has recurred in a myriad of contrasting settings, from public health to urban design. Readers will find a resonance here with materialist perspectives on urban history and landscape iconography through a recurring emphasis on the crucial role of capital in the shaping of urban space. In some instances the role of capital takes the form of municipal bonds or other forms of investment in the urban environment; in other cases nature is itself transformed into new commodities such as urban parks based on the sophisticated interaction between real estate speculation and landscape aesthetics.[6] To argue that capital is a key dimension to the production of an urbanized and commodified nature in New York City is not to deny the influence of other factors but simply to place political considerations at the center of our analysis. The production of urban nature not only involves the transformation of capital but simultaneously intersects with the changing role of the state, emerging metropolitan cultures of nature, and wider shifts in the social and political complexion of city life. The complex cultural hybridities that characterize the development and experience of urban nature have a material basis not only in the city itself but also in the processes of social and economic transformation that have enabled New York City to gain its ascendancy as a global city.

The terms "nature," "landscape," and "environment" are used loosely here so as not to exclude different realms of urban nature. Facile distinctions between landscape and "cityscape," for instance, are rejected in order to counter the misleading emphasis on rural-urban distinctions in much existing literature.[7] A broad and inclusive definition of landscape allows the urban experience to be explored in relation to changing conceptions of nature without separating the technical, political, and aesthetic dimensions of urban space. Much discussion about cities adopts a conception of urban change as little more than the conscious design of a few individuals: the oscillation between architect, engineer, and planner is replicated in an urban troika of buildings, infrastructure, and urban design. The city is conceived as no more than an agglomeration of its parts derived from the fragmentary Cartesian-Lockean world view transposed to the analysis of urban form. Yet two critical dimensions are missing from this formulation: history and nature. History is central to our analysis because the relationships between different facets of urban space change over time: consider, for example, the ebb and flow of different constellations of social power; the fluctuating economic context for capital investment; and the development of distinctive forms of urban consciousness. The transformation of nature provides the other missing element: in the absence of parks or green spaces, the city appears oppressive and dehumanized; the construction of elaborate urban infrastructures has largely dispelled the health-threatening environments of the past; and the cultural resources of nature have been woven into different forms of capital in urban space.

Even the city itself is from an analytical point of view an abstraction: a physical or cultural motif that can easily become divorced from any wider consideration of the dynamics of urban change. The morphological boundaries of the city become a kind of analytical prison that serves to delimit our understanding of the urban process. The city can in other words become fetishized as something "natural" that has evolved separately from its social and historical context. The naturalization of social power is one of the most powerful ways in which nature and ideology intersect. The word "ideology" is used here to denote a dominant imprint of ideas on social life that sustains existing power relations in society; it is a constellation of cultural and political power that seeps through the pores of everyday life. Ideology is a hidden language of power that extends from the design of parks to the social stigma attached to dirty neighborhoods.[8] Yet in invoking this idea of nature as a mask that hides the "real" mechanisms operating in urban space, the book does not seek to develop a crudely materialist interpretation of the urban process: the focus here is on the mutually constitutive relations between nature as biophysical fabric and the symbolic power of nature as a cultural representation of imaginary landscapes.

The processes of modernization and urbanization in nineteenth-century America developed in a state of tension with different conceptions of nature and landscape in the cultural and political imagination. The growth of cities not only transformed ever greater swaths of the "frontier" landscape to meet urban needs but also fostered the emergence of new kinds of cultural interactions with nature. In an American context we can trace antiurban sentiment to the very different intellectual lineages of Thomas Jefferson and Henry Thoreau. For Jefferson, the agrarian idyll of small-town America fostered greater human virtue, whereas for Thoreau the beauty of nature was a solitary and transcendental experience rooted in romantic idealism. Jefferson feared that cities were "pestilential to the morals, the health, and the liberties of man"; urbanization was in other words antithetical to what he saw as the essence of American values.[9] In contrast, Thoreau projected the privileged voice of a metropolitan elite onto an imaginary "first nature" that would not have been widely recognized by either early settlers or Native Americans.[10] The idea of a radical separation between nature and cities is a powerful current running through Western environmental thought. Yet this city-country divide is at root ideological rather than analytical. We are presented with a series of false comparisons rather than substantive observations on the nature of landscape change and the dynamics of capitalist urbanization.[11]

Metaphors drawn from nature and the natural sciences have had a longstanding impact on urban thought. In the seventeenth century, for example, new conceptions of economic exchange and medical science emphasizing movement, mobility, and circulation began to shape Enlightenment perspectives on urban form. The developments in medical knowledge advanced by William Harvey, Thomas Willis, and other pioneers of the empirical sciences fed into new sanitary discourses focusing on the free movement of water, air, and citizens through the body of the city.[12] Organic metaphors describing the "circulatory health" of fast-growing nineteenth-century cities became ideologically explicit when used to characterize the new and frightening forms of social polarization, political instability, and economic uncertainty that these cities experienced. In the twentieth century, a range of technological advances facilitated a new mediation between organic metaphors and the production of urban space. In 1965, for example, the engineer Abel Wolman outlined "the complete metabolism of the modern city" as the culmination of advances in the technical organization of urban space.[13] Yet these metabolic metaphors treat the city as a discrete physical entity. The "body of the city" is considered in isolation from the wider determinants of urban form, and the social production of space is downplayed in relation to the technical mastery of cities. For Wolman, the task of the engineer is to ensure the smooth functioning of the urban system and to apply the latest advances in technical knowledge to the resolution of urban problems. Healthy urban metabolism rests on the interaction between the constituent elements of the urban process derived from water, energy, and other raw materials. The digestive and alimentary dimensions of the modern city are derived from an array of inputs and outputs that must be managed by a team of technical experts. Even where the spatial and conceptual parameters of organic metaphors have been extended, as in the rhizomatic "space of flows" model developed by Gilles Deleuze and Félix Guattari, we find little sense of any distinction between the relative significance of different movements in urban space: the most useful dimensions to their perspective are the exposure of the morphological perimeters of the city as an arbitrary division within a wider system of flows.[14] Above all, however, metabolic conceptions of urban form tend to neglect the flow of capital. It is capital that represents the most powerful circulatory dynamic in the production of modern cities, yet its presence remains obscured in most technical or libidinal accounts of the urban process.

In contrast to the space of flows model, the concept of "cyborg urbanization" allows us to articulate a dialectically conceived version of urban metabolism relating technical developments to a broader cultural and political terrain. If we take the cyborg to be "a cybernetic organism, a hybrid of machine and organism," then urban infrastructures can be conceptualized as a vast life support system.[15] The architectural historian Alberto Abriani, for example, describes how the development of plumbing technologies in the modern era transformed the home into "an inseparable part of the urban body: the individual organ (the home network) becomes a member of the social body (the city's public network)."[16] The notion of "cyborg urbanization" is a useful way of extending existing conceptions of nature in cities by emphasizing the physical vulnerability of the human body as part of a hierarchy of larger-scale social and metabolic systems. The cyborg metaphor reveals the interaction between social and biophysical processes that produce urban space and sustain the possibilities for everyday life in the modern city. Above all, this hybrid perspective illuminates the tension between the city as an abstract arena for capital and as a lived space for human interaction and cultural meaning.

In contrast to the cyborg metaphor, most conceptions of "ecological urbanism" drawn from the environmental literature rest on a sharp delineation between the "natural" and the "artificial." Over the last twenty years there has been growing interest in the "ecological city" as an alternative to the environmentally destructive, violent, and socially divisive characteristics of contemporary urbanism. The "ecological city" draws together a series of interrelated themes: the use of nature as a "blueprint" or set of rules for the organization of human society; an organic conception of the regional economy as a largely self-contained form of social organization; an aesthetic predilection for the "urban pastoral" rooted in nineteenth-century romanticism and a belief in the "curative" and "regenerative" powers of nature; and an elision between the ecological world view and a wider critique of modernist thought and design. The landscape architect Ian McHarg, for example, longs for a city that emulates "an ecosystem in dynamic equilibrium, stable and complex."[17] But how do we reconcile the analogy of ecological stability with the prospect for social change? For McHarg, the "rules" of urban design are to be found in nature, since ecological science provides "not only an explanation, but also a command."[18] The architects Natasha Nicholson and Pamela Charlick extend the ecological analogy even further by presenting the science of complexity as a new urban paradigm that ranges from the biophysical realm to the economic structure of the city. Yet in merging aesthetic and scientific themes, this antimodern manifesto for the ecological city overlooks the continuities in capitalist urbanization that continue to shape the dynamics of urban change. Nicholson and Charlick contrast the perceived linearity of the modernist city with a homeostatic conception of the ecological city as a "self-organizing living organism," where "the individual and the whole maintain their integrity in a co-operative and non-hierarchical organization."[19] Their Gaian urbanism conceives of the city as a single organism and is predicated on a sharp distinction between the modern and the postmodern. It seems that the terms "ecological" and "natural" have become interchanged in a design sense to produce a kind of ideological tautology that rests on the authority of science to elucidate both the mechanisms and appropriate outcomes of landscape change.

Yet the ecological view is critically deficient with respect to the social production of nature. At root, the ecological perspective does not question the role of capital in the production of urban space and is largely silent on questions of social power. Despite the claims of "newness," the ecological fusion of nature and society does not represent a radical break with the past but simply reworks the long-standing Enlightenment preoccupation with the unification of Nature and Reason. In 1765, for example, Marc-Antoine Laugier's treatise on architectural theory likened urban form to a natural phenomenon. The processes of urban transformation unleashed in the protocapitalist city were conceived as part of a natural process conforming to universal laws of growth and change in nature.[20] Similarly, Francesco Milizia, writing in 1813, likened the city to a "forest" derived from a medley of organic forms in which "the streets must radiate starlike, there like a goose-foot, on the one side in herringbone pattern, on the other like a fan."[21] What is of particular interest in an American context, however, is the early emergence of the grid plan pioneered by Pierre L'Enfant in Washington and

William Penn in Philadelphia, which made the pragmatic underpinnings of capitalist urban design far more explicit than in their European counterparts.[22] It is this tension between naturalized forms of urban design and the abstract demands of capital in urban space that lends such poignancy to Central Park, for example, as the "rectilinear eye of the storm" in the Manhattan real estate market.[23] Fragments of nineteenth-century urban nature provide powerful material and symbolic links between the aesthetic discourses of the past and the contemporary promotion of organic templates for ecological stability and social cohesion.

The combination of ecological ideas with urban analysis is fraught with difficulty. If we want to incorporate the independent agency of nature into our analysis, we need to be sensitive to the way in which biophysical processes are mediated through human cultures: explanation in the physical and biological sciences is rooted in metaphors that are social and cultural in origin, even if the phenomena under investigation have an ontological status of their own. To call for an ecologically based urbanism is to replace the historical analysis of social change with an arbitrary alternative. The environmentalist Herbert Girardet, for example, suggests that cities must change from a "parasitic" to a "symbiotic" relationship with their host environment in order to prevent widespread social and environmental calamity in the twenty-first century.[24] This perspective is not only misplaced in terms of its analysis of the urban process but also suffused with a disciplinarian response to social and spatial disorder. The implication is that an autarchic and inward-looking urbanism will replace the modern metropolis in order to sustain a different set of relations between society and nature. The difficulty with such formulations is that the ecological template for urban change is always assumed to lie outside of society itself. The recent popularization of the "ecological footprint" as a metaphor for understanding the environmental impact of cities is similarly trapped within a bioregional conception of urban and regional planning that asserts that cities are antithetical to "natural material cycles."[25] The aggregate ecological impact of conurbations is portrayed as the focal point for any understanding of the global dislocation between social and ecological systems. Yet there is no analysis of how the patterns of resource use and ecological degradation are underpinned by social and economic relations that operate in a different way from biophysical systems. There is, in other words, an epistemological blurring of social and ecological systems that allows concepts such as "carrying capacity" and "population overshoot" to be switched at will between markedly different contexts. Where ecological perspectives are combined with more radical conceptions of social change, we find a more productive interaction between insights derived from the social and environmental sciences. The political scientist Thomas Jahn, for example, has developed a series of imaginative metaphors for conceptualizing nature-culture relations in urban space. Jahn's model of "social ecology" dispenses with the artificial city-country distinction and builds an integrated approach to the interpretation of regional environmental change.[26] Yet even the most politically engaged variants of social ecology, epitomized by the writings of Murray Bookchin, harbor a deep ambivalence toward the city and its "unnatural" scale in relation to more communitarian forms of human existence.[27] We must look to other intellectual traditions to uncover more productive interactions between different cultures of nature and progressive dimensions to urban politics.

The sophisticated cultural histories of nature developed by scholars such as William Cronon, Donna Haraway, and Alexander Wilson have sought to explore the dynamics of environmental thought from the assumption that we are inside rather than outside of nature. William Cronon has examined the intersection between social and biophysical processes in new ways in order to extend the scope of environmental history to include the insights of anthropologists, geographers, and ecologists. He has explored the distinctive kinds of property relations that have underpinned the transformation of cities and landscapes in the modern era.[28] The critical investigations of Donna Haraway have similarly extended our understanding of the relationship between nature and society. She has examined the relationship between scientific knowledge and social power to show how nature and culture are combined in a myriad of different ways. In place of the disembodied scholar who works from a Euclidean geometry of visual metaphors, she suggests the rather different ocular metaphor of light diffracted through a prism to produce an array of potential insights. Haraway advocates a move beyond the omniscient vantage points of scientific models rooted in the simplistic mimesis of social reality toward a subtle and multiperspectival purchase on our field of investigation.[29] With the writings of Alexander Wilson we find a wealth of insights into the most prosaic of everyday places and experiences. Wilson develops a cultural history of nature with great sensitivity toward the material and iconographic dimensions of the human landscape. He teases out the changing meaning of nature from a tight knot of political and philosophical assumptions. He questions, for example, the reductionist view that nature is something inherently good, which has suffused American environmental thought from Henry Thoreau to Rachel Carson.[30] Our cultural responses to nature are never self-evident but are derived from a complex medley of often contradictory ideas. It is because of the innate ambiguity of nature and its ability to provide ideological veneer to almost any argument that we need to critically examine how changing meanings of nature have intersected with wider debates about urban change.

In order to make better sense of the production of urban nature, it is necessary to consider how different cultures of nature have been created through an interrelated set of transformations in the modern era. Modernity involved the dissection of urban space and the recombination of different elements in the service of a complex set of political, economic, and cultural ideals. The chaotic and disease-ridden nineteenth-century city was transformed into what the Belgian surrealist Paul Delvaux termed an "empire of lights" in which the progressive forces of urban society could flourish and develop under the glare of new technologies.[31] But there was always an ambiguity running through modernist visions of urban space marked by a blurring of the boundary between technical rationalization and social control. A multiplicity of cultural and philosophical modernisms were hidden behind the force of modernization and urban change. Reformist impulses evolved in an uneasy symbiosis with the puritanical and authoritarian preoccupations of nineteenth-century social and political thought. Changing technologies and social attitudes evolved alongside sharpened gender differentiations and new cultures of health and domesticity. Emerging cultures of professional expertise conflicted with the popular dynamics of working-class suffrage and the growing powers of municipal governance. Within this shifting patchwork of change we find that the use and meaning of nature were altered to meet new standards of hygiene and material sustenance, to contribute to the physical fabric of urban design, and to underpin the sensory realignment of everyday life. Nature was not only reworked in a myriad of ways to reflect new advances in science, engineering, and other fields, but was also represented back to urban societies in an increasingly sophisticated array of cultural forms ranging from art and literature to the latest innovations in landscape design.[32]

The early decades of the twentieth century saw the acme of "scientific urbanism," marked by the advent of technological modernism and regional planning. Twentieth-century developments such as the increasing technical rationalization of urban space and the decisive impact of public works projects form part of a longer-term continuity in patterns of urban change, linked to early public health reform and to the gradual ascendancy of professional expertise in urban governance. The architectural historian Alberto Abriani suggests that "the parable of the sanitary engineer" lasts about fifty years, from 1870 to 1920.[33] But Abriani's periodization is too abrupt: the zenith of the technological mastery of urban space surely came in the late 1950s and early 1960s, when advances in systems theory and urban algorithms were combined with the technical and economic impetus of post-World War II urban renewal.[34] The postwar promise of urban reconstruction, the advance of suburbia, and the pervasive allure of the "auto-house-electrical appliance complex" appeared to open up a new vista of democratic urbanism across Europe and North America.[35] In many respects, of course, the roots of dispersed urbanism and the rise of suburbia can be traced back to the nineteenth century, as the early parkways and model settlements emerged as an antidote to the dismal environs of the industrial city. The economic logic of postwar urban decline was simply the manifestation of incipient trends contained within nineteenth-century urban design traditions, which could now be radically extended through the combined pressures of economic prosperity, cultural aspiration, and technological innovation.

It is from the 1960s onward that we find a systematic reworking of the "municipal managerialist" ethos in urban space. The relationship between urban governance, capital, and the built environment was recast. The tension between political and economic elites and the urban grass roots was thrown open to much wider scrutiny, marked by an exhaustion of existing paradigms associated with technical mastery and control over urban systems. For perhaps the first time in the history of modern urbanism the place of nature within the public realm was opened up to competing conceptions of urban design that extended beyond the elite cabals of aesthetic judgment. The elision between the social and technical dimensions of urban space was no longer politically credible. The historian André Guillerme describes how engineering science now faced its "chilly limits" as its sources of fiscal and ideological sustenance evaporated.[36] Many symbols of modernity such as elaborate infrastructure networks, bridges, tunnels, and public spaces had fallen into disrepair, their economic use and cultural meaning altered in an unfamiliar landscape dominated by new patterns of urban development. The omniscient perspective of "total planning" had become both a technical and a philosophical anachronism. The dynamics of urban decline fatally damaged the utopian ideals of Le Corbusier, Eugène Hénard, and the urban futurists, with their faith in decentralized, spatially segregated, and technologically driven urban form. The rejection of modernist urbanism marked a resistance to the intrusion of technical rationality into what Jürgen Habermas terms the "life world" but also masked the diversity of strands to twentieth-century architecture and urban design.[37]

The political turmoil of the 1960s in Europe and North America set in train a complex set of changes in relations between the "public interest" (as expressed through various government agencies) and alternative perspectives on urban form. The notion of public space became irrevocably problematized in relation to what Jonathan Raban refers to as "the hypothetical consensus of urban life."[38] Classic conceptions of the bourgeois public sphere now conflicted with newly emerging social and political realities that placed much greater emphasis on difference, diversity, and market-driven dynamics of cultural change. The very idea of a "public landscape" was now subject to intense critical scrutiny, leading to open conflict between "avant-garde" and "populist" conceptions of urban design.[39] At the same time, however, the "revanchist city" with its recycling of historical iconographies in the service of a highly commodified and increasingly privatized public realm has worked to create new forms of social and spatial exclusion.[40] The "social cleansing" of the historically significant core areas of cities, the policy of zero tolerance toward social deviance, and the proliferation of new forms of "public-private" partnership have spawned a disciplinarian geography of controlled leisure and commodified homogeneity in which the tax base of cities has been diverted from social welfare to corporate subsidy. Inequalities in the experience of the urban environment have widened through disparate contexts ranging from the plight of the homeless and the "new poor" to the development of private atria and increasingly exclusive conceptions of the design and management of public space, as part of a striking synergy between nineteenth-century moralism and contemporary urbanism.

Though we can discern some agreement surrounding the characteristics of the modern metropolis, we cannot find a similar degree of consensus on the nature of contemporary urbanism. The literature is overflowing with different perspectives, ranging from "gulag urbanism," "edge cities," and dystopian future dioramas to celebrations of the postmodern "heteropolis" and the growing significance of "virtual spaces." Urban scholars have analyzed new kinds of relationships between economic and cultural developments; changing flows of urban investment and their architectural expression; intensified forms of social exclusion and spatial polarization; new forms of municipal governance and civic boosterism; and the shift toward more flexible and informal patterns of economic activity. "Postmodern urbanism," for want of a better expression, usefully captures a range of political, economic, and cultural developments that have swept through global cities in the last two decades of the twentieth century.[41] More profoundly, however, the postmodernity debate has challenged us to rethink how we understand urban spaces; it has opened up the city to a range of other voices long suppressed within the "expert vision" inherited from the technocratic legacies of urban planning and functionalist accounts of urban form. The search for total explanation, whether in a positivist or a structuralist guise, has been supplanted by recognition of the diverse, partial, and embodied dimensions to urban knowledge. Although the term "postmodern" has been repeatedly splintered and refracted, it continues to serve as a useful focal point for discussing the contours of contemporary urbanism. Yet at the same time the postmodernity debate has been fundamentally misleading. The poststructuralist impetus running through much recent urban scholarship has failed to elucidate the substantive connections between material and cultural transformations of cities. A focus on the social construction of space has fostered a pervasive dualism between the "real" and the "imaginary" city rooted in an analytical separation between cultural and economic change.[42]

We are living through a dislocation from the metropolitan nature forged under nineteenth-century urbanism with its characteristic modes of consumption and technical expertise. The place of nature within cities has been radically reconfigured in response to these changes, through new kinds of relations between nature and capital, between private and public spaces, and among competing perspectives on the meaning and iconographic significance of urban landscapes. Though the modernist legacy of "urban perfection" has not disappeared, its critical impulse has been radically diluted. To borrow Kevin Lynch's terminology from the early 1960s, we are confronted by a problem of "legibility" in the urban landscape, which if anything has become heightened since the time Lynch was writing.[43] The sheer scale and complexity of the contemporary city pose fundamental dilemmas about the comprehensibility of urban space either as visual diorama or as a substantive focus of analysis. The last twenty years have seen a divergence between the critical acuity of urban scholarship and the urban *Realpolitik* of the post-Fordist era. The challenge is to build a conception of urban nature that is sensitive to the social and historical contexts that produce the built environment and imbue places with cultural meaning.

Our story begins with the provision of water for New York City. Chapter 1 traces the context for the development of the city's first comprehensive waterworks and the gradual movement toward the plumbing of an entire city through a centralized system of technological and organizational control. We follow a period of major infrastructure investment from the mid-nineteenth century until the 1960s associated with the completion of a vast system of aqueducts, dams, and reservoirs. Since the 1970s, however, a long phase of relative stability in urban water provision has become unraveled; the "total vision" of water resources management inherited from the Progressive era, and its technological, organizational, and bacteriological antecedents in the nineteenth century, has been displaced by an aura of uncertainty. Chapter 2 focuses on the creation of Central Park in order to emphasize a series of ambiguities running through different conceptions of public space, the public realm, and the creation of a symbolic urban order. The recycling of historical motifs in the service of current real estate values provides a powerful link between nineteenth-century discourses of urban beautification and the contemporary search for nature-based urban visions inspired by Olmstedian conceptions of social and ecological order. Chapter 3 examines the development of landscaped roads or parkways. We consider the intersection between modernist aesthetics and technology that produced innovative new dimensions to the urban landscape; the relationship between road building and technological modernism is explored as the outcome of a combination of political, economic, and cultural developments that underpinned the transformation of the New York metropolitan region. In chapter 4 the grassroots city of the urban ghetto is explored with a study of radical politics in the Puerto Rican *barrio* which emerged out of the civil rights movement of the 1960s. This example develops our understanding of urban nature in two main ways. First, it links the transformation of the New York environment to the colonial legacies of US imperialism in the Caribbean as part of a geopolitical ecological frontier that goes far beyond the familiar metaphors of regional hydrological cycles or patterns of timber and food production within North America. Second, the radical intersections between race, class, and gender pursued through the politics of the ghetto extend our understanding of urban nature to encompass the politics of the body and the human environment. Our last visit to the New York metropolis examines the mobilization of a working-class neighborhood against pollution in the 1990s. A new geography of power has emerged in which the city's ecological hinterland has been extended into the impoverished rural communities of the South as part of a growing trade in urban detritus.

1

Water, Space, and Power

New York City is the most thirsty of all great cities.

—*Jean Gottmann*[1]

The provision of water for New York City is one of the most elaborate feats of civil engineering in the history of North American urbanization. As the city grew, it extended an "ecological frontier" of water technologies deep into upstate New York. The city's modern water supply system, which has been intermittently under construction since the 1830s, now extends across the largest water catchment area in the United States, collecting water from nearly 2,000 square miles of sparsely populated mountains, lakes, and forests of the Catskill region along with the smaller and more densely populated Croton catchment in closer proximity to the city. The city's elaborate water infrastructure now includes 19 collecting reservoirs, two city water tunnels, the world's largest storage tanks, and nearly 6,000 miles of gravity-fed water mains. This vast network delivers 1.3 billion gallons of water a day to 9 million people (see figures 1.1 and 1.2).

Some sense of the scale and complexity of the water supply system can be illustrated by the city's new six-billion-dollar water tunnel, City Water Tunnel No. 3, one of the biggest civil engineering projects in America, which was conceived in 1954 and has been under construction since 1970. Although the project

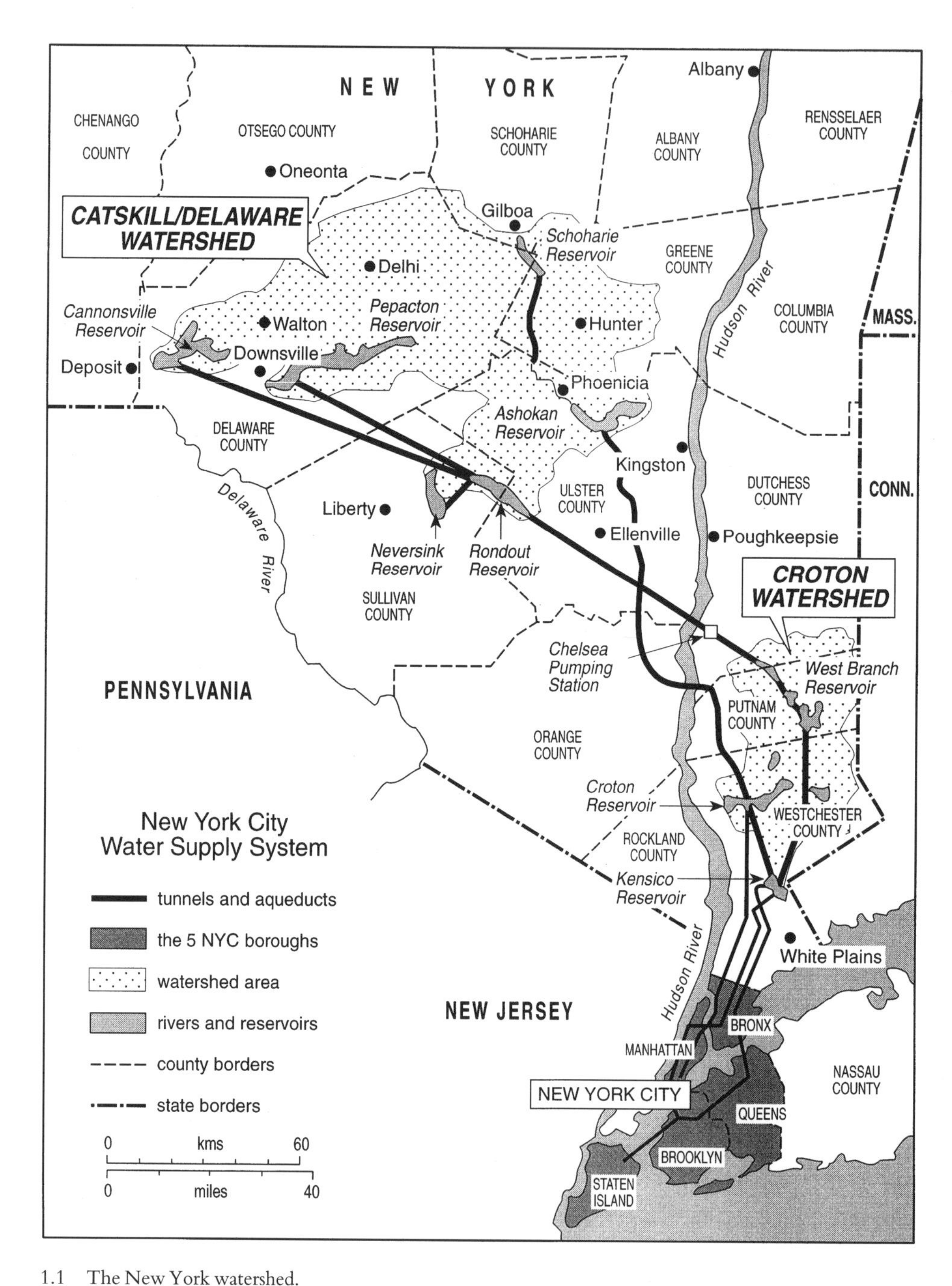

1.1 The New York watershed.
Source: New York City Department of Environmental Protection.

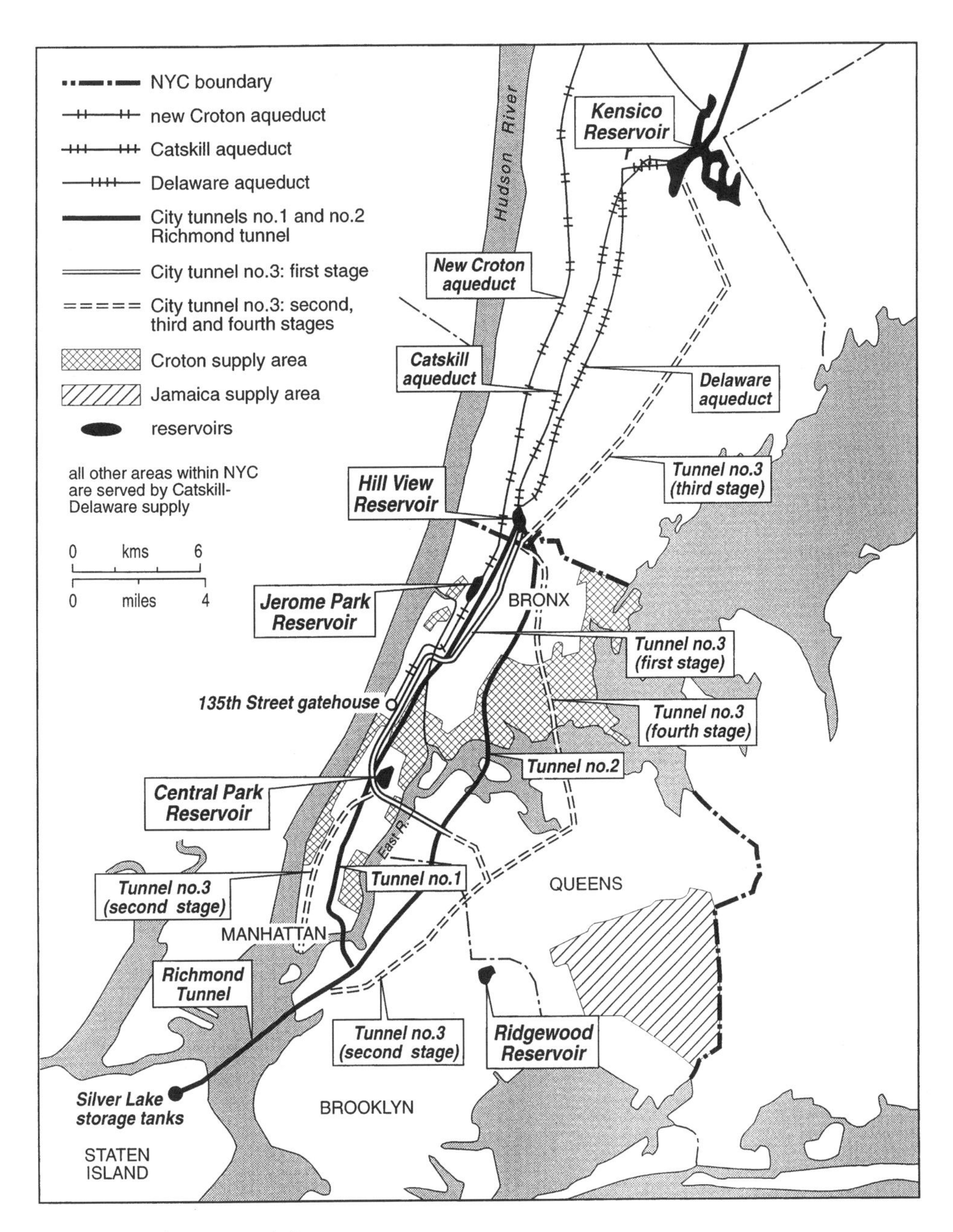

1.2 New York City water infrastructure.
Source: New York City Department of Environmental Protection.

was halted and nearly abandoned during the city's fiscal crisis of the mid-1970s, the first section was finally opened in 1998, and the whole tunnel is expected to be completed by 2020. Beneath an inconspicuous door in Van Cortlandt Park in the Bronx lies a 250-foot shaft linked to a dazzling subterranean valve chamber over 600 feet long that connects the city to its upstate sources of water supply. This extraordinary engineering achievement has an austere, utilitarian aesthetic reminiscent of the most impressive American water technologies of the twentieth century, such as Gordon B. Kauffmann's Hoover Dam and Eero Saarinen's Watersphere. At the tunnel's official opening in August 1998, a giant fountain was activated in the Central Park Reservoir that had first been used in 1917 to celebrate the arrival of water from the Catskill Mountains.[2]

The history of cities can be read as a history of water. The historian Nelson Blake contends that "the indispensable precondition to the great growth of American cities during the nineteenth century was a recognition of the vital importance of water supply."[3] In the absence of plentiful supplies of water, cities are faced with the threats of fire, disease, social unrest, and material impoverishment. To trace the flow of water through cities is to illuminate the functioning of modern societies in all their complexity. Water is a multiple entity: it possesses its own biophysical laws and properties, but in its interaction with human societies it is simultaneously shaped by political, cultural, and scientific factors. For the historian Jean-Pierre Goubert, the modern era has seen a series of transformations in the use and meaning of water. The premodern waters of myth and salvation have been "subjugated, domesticated, mechanized and made profitable."[4] During the nineteenth century the water technology of the cities of Europe and North America evolved from an organic form, in which limited flows of water were combined with the harvesting of human wastes as fertilizers, into a modern hydrological structure in which far greater quantities of water were transported through the city in an ever more complex network of pipes and sewers. The control of water in the modern city now extends from the regional hydrological cycle to the application of plumbing technologies within the home. Water, in its multiple uses and transformations, flows between these different kinds of urban spaces linking diverse elements such as capital markets and domestic technologies into a multitiered social reality.

The story of urban water supply is a powerful element in the field of environmental history. Scholars have grappled with this dynamic from a variety of vantage points. In the American West, for example, the interaction between water and cities has become woven into a powerful narrative of the technological ingenuity behind urban growth. In New York, however, the relationship between nature and the urban landscape is far less clear. The architectural critic and landscape designer William Morrish, for example, suggests that the "dominant contextual elements of Los Angeles are not architectural, but natural." He describes how "the mountains surrounding Los Angeles are viewed as part of the formal vocabulary of the urban landscape," whereas New York, by contrast, is dominated by its architecture rather than its physical setting.[5] This chapter will show, however, that this comparison is misleading in ecological if not visual terms, since New York, like Los Angeles, is woven into its mountainous hinterland by an elaborate network of water technologies in order to transport billions of gallons of fresh water.

The landscape of upstate New York has been sculpted into a life-sustaining circulatory system through the interaction of the flow of water and the flow of money. Yet this double circulation of water and money is easily overlooked. The more distant parts of the city's watershed still resemble a Thoreauvian wilderness: one can trek through parts of the Catskill Mountains without encountering another human being. It is easy to imagine that you have entered a fragment of primal nature, but there are signs of human influence all around: the absence of large mammals; the patches of trees that have regenerated since the land was logged in the nineteenth century; and the network of paths that bear witness to many centuries of intense human activity. Most remarkable of all, however, is the fact that you are standing inside New York City's water system. The hydrological cycle of a whole region has been harnessed to provide water for a city: the rain dripping down through the leaves of hemlock trees will eventually find its way into the pipes and taps of millions of homes.

1.1 WATER AND THE NASCENT CIVIC REALM

Access to water was a constant problem for early settlers as they began to colonize the southern tip of Manhattan Island during the seventeenth century. In New Amsterdam, as the settlement was then known, there were frequent water shortages. Although many businesses and wealthier dwellings had their own wells, those without private water sources, principally the urban poor, depended on public wells, the first recorded of which was dug at what is now the intersection of Bowling Green and Broadway in 1658. The meager supplies from wells were supplemented by the collection of rainwater through the extensive use of cisterns made of masonry or wood.[6] With the growth of the city many of the wells became contaminated by privies, cesspools, and the drainage of dirty water from the streets. By the mid-eighteenth century the Swedish traveler and diarist Peter Kalm reported that even travelers' horses were reluctant to drink New York water. "There is no good water in the town itself," writes Kalm. "This want of good water is hard on strangers' horses that come to the place, for they do not like to drink the well water." [7] Records suggest that the only major source of good water remaining in the eighteenth century was derived from a spring outside the city called the Collect Pond in what is now Park Row. Water derived from the Collect Pond was sold to those who could afford it by water vendors known as "Tea-water Men." By 1785, however, the city's main source of water had degenerated into what the *New York Journal* described as "a very sink and common sewer."[8] In 1808 the city resolved to fill in the stinking Collect Pond in order to provide employment for some of the thousands of sailors and laborers thrown out of work by the Embargo Act. This early large-scale public works project reveals from the outset how political and economic factors would predominate in all discussions surrounding water supply and the construction of urban infrastructure. Rising land values merely accelerated the rate of development around the Collect Pond, and by 1815 the site had been completely filled in.[9]

Crowded, insanitary conditions in New York led to repeated epidemics of infectious disease. A series of outbreaks of yellow fever, a deadly virus carried by the mosquito *Aedes aegypti,* was recorded during the eighteenth century. Between 1791 and 1821 yellow fever epidemics became even more frequent. The most severe outbreak was recorded in 1798, when we find some of the earliest direct links made by city physicians between poor sanitation and disease. In the 1805 outbreak nearly half the city's population fled and the economic survival of the port of New York was threatened.[10] As for cholera, the great scourge of the nineteenth-century city, serious epidemics struck in 1832, 1849, 1854, 1866, and 1892 (the last major outbreak). The 1832 cholera outbreak, in which more than three thousand people died, was to prove pivotal in the early politicization of public health reform.[11] "On no former occasion," wrote the physician John W. Francis, "has New-York, frequently visited by the direful ravages of the yellow-fever, exhibited a more melancholy spectacle. Of a resident population of two hundred and twenty thousand . . . at least one-third are now dispersed in every direction."[12] Yet for Francis, as for most of his contemporaries, the cause of cholera remained inexplicable:

> It is conceded by all, that the origin of epidemic diseases is still enveloped in great obscurity; and the theories on this subject, whether referring to a distempered state of the atmosphere, to exhalations from putrid animal or vegetable matter, or to specific contagion, have been alike conjectural and unsatisfactory. The cholera, like all preceding epidemics, has exercised, but without any very useful results, the ingenuity of the speculative and philosophical observer.[13]

Francis described how cholera extended to "the innumerable circumstances connected with the economy of man in every state and condition" and attacked "rich and poor, native and stranger, young and old."[14] Despite the indiscriminate pattern of cholera morbidity across different social classes, the moralistic and superstitious responses to disease were only partially displaced by emerging concerns with urban sanitation. In the mid-nineteenth century we find a medley of perspectives on the cause of disease, combined to produce a moral geography of illness. In 1849, for example, the New York City Board of Health could confidently assert that "the general cause of the disease appears to exist in the atmosphere" and that "the agency of various exciting causes is generally necessary to develop the disease. Among these causes the principal are the existence of filth and imperfect ventilation, irregularities and imprudencies in the mode of living, and mental disturbance."[15]

In the prebacteriological era the problem of water supply was predominantly perceived to be one of bad taste and insufficient quantity; the danger and nuisance of poor water quality was not widely linked to disease. The cholera epidemics did, however, begin to widen and intensify public debate over urban sanitation, even if the precise mechanisms of contagion were imperfectly understood.[16] Commentators such as the physician Martyn Paine noted that the prevalence of cholera was worse in Paris and Montreal than in New York and began to develop a more rigorous analysis of variations in ventilation and cleanliness that might contribute to the severity of the disease.[17] The cholera epidemics of the nineteenth century were, above all, a transitional moment in the history of capitalist urbanization, as new trade routes exposed insanitary cities to the threat of disease before there were any concomitant advances in the science of epidemiology or the practice of public health. They were, in other words, the outcome of an uneven modernity that exposed a series of political, cultural, and scientific contradictions running through urban society. While it is undoubtedly the case that changing attitudes toward public health were a significant pretext for the modernization of the city's water supply (and environmental conditions more generally), these reforming impulses belonged to a wider political and economic context for capital investment in the physical infrastructure of cities. These changes rested on an emerging commonality of interests between the power of technical elites and the economic logic behind the reordering of urban space.

The late eighteenth and early nineteenth centuries were the worst period of all in the history of the city's water supply and heralded an increasingly panicky series of investigations into alternative sources. In 1774 the city's Common Council approved a plan put forward by the Irish-born engineer Christopher Colles to build the city's first municipal waterworks using a hilltop pump from which water could flow in any direction by gravity.[18] By approving Colles's scheme, the city embarked on a completely new approach involving the construction of a much more sophisticated water supply system than the existing wells and ponds. Construction began in 1774 using a steam engine designed by Colles himself, connected to a network of water pipes made from pine wood. The city financed the construction by issuing 5 percent bonds known as "water works money." In the event, the project was destroyed before its completion during the British occupation of the city in 1776. In spite of this setback, the city was slowly but inexorably widening its responsibilities for water supply. In 1792, for example, it began to use tax revenues for the digging of new wells, so that by 1809 there was a network of 249 public wells.[19]

Before any collective solution could be found for the city's water supply problem, however, there was an extraordinary historical detour created by rival attempts to control New York's banking system. In the 1790s the only banks in New York were controlled by the Federalists under Alexander Hamilton. The leading Republican, Aaron Burr, knew that his attempts to set up a rival bank would be thwarted by his political opponents in the State Legislature. Burr successfully lobbied against the construction of a public water system on the grounds that the State Legislature would be unwilling or unable to adequately finance it. In 1799 he then succeeded in passing a bill that granted a charter to a little-known water company called the Manhattan Water Company of which he was chair of the board. Hidden in the redrafted bill was a clause that allowed the company to use its surplus capital "in the purchase of public or other stock, or in any other moneyed transactions or operations."[20] Having set up the Manhattan Water Company, Burr rejected the more expensive options contained in the company's charter of diverting clean water from Westchester County and the Bronx River (which would have used up too much of the company's capital) and opted instead to drill a well adjacent to the polluted Collect Pond with cheap wooden pipes. The company never constructed the steam pump and million-gallon reservoir that it had promised the city but relied instead on horse power in combination with a small reservoir one-tenth the size, adorned with a "false front of four Doric columns supporting a recumbent figure of Oceanus."[21] As well as being of poor quality, the water was also extremely expensive at 20 dollars a year, leaving most of the city reliant on rainwater collected in rooftop cisterns or "buying an occasional pailful from fetid wells."[22] Throughout its entire operations the Manhattan Company never laid more than twenty-five miles of water mains, but the profits were used to set up the Bank of Manhattan Company (later to become Chase Manhattan after the 1955 merger with Chase National Bank). So inadequate was the company's charter, which had been approved by the unwary State Legislature, that the city was forced to use its own revenues for flushing gutters, piping water to markets, and repairing streets after the laying of pipes. The city's attempt to buy the waterworks of the Manhattan Company in 1808 met with public indignation: not only had the company operated under the minimum legal obligations of its charter but it now stood to benefit financially from its own failure. In 1822 there was yet another serious outbreak of yellow fever and fifteen prominent physicians signed a certificate warning that the Manhattan Company's water was unfit for human consumption. By 1830 the Manhattan Company still served no more than a third of the city: the rest of the population, who now numbered over 200,000, remained dependent on polluted wells or were forced to buy water from private vendors at exorbitant prices. The "Tea-water Men" and other water vendors were earning about $275,000 a year; ships were paying $50,000 a year to have their casks filled; and fire was destroying around a quarter of a million dollars' worth of property a year.[23]

Under the twin impetus of industrialization and immigration, New York was now the largest and fastest-growing city in America. A building boom in the 1820s and 1830s spread rapidly northward from the southern tip of Manhattan Island as former residential districts became converted into more lucrative commercial and industrial premises. Beyond the business district, with its dense concentration of new urban infrastructures such as sewers and gas lighting, the rest of the city languished in a state of anarchy. In the early 1830s public interventions multiplied with the construction of a new wooden reservoir on 14th Street and the laying of extra pipes in the driest parts of the city, but these efforts amounted to little more than tinkering at the edges of the problem.

During the early decades of the nineteenth century we can observe the emergence of a shift in the politics of water which began to undermine the claims of private provision. This set in train a reformist urban vision based on a more sophisticated conception of the technical, administrative, and financial dimensions to public works. We should be careful, however, not to overstate the significance of public health concerns in relation to the broader political and economic dynamics behind the modernization of urban infrastructure. In 1798, for example, the engineer Joseph Browne had called for the extension of waterworks to new sources beyond the city's political boundaries, but the continuing intransigence of city authorities in the early nineteenth century stemmed from a reluctance to finance any program of public works that might significantly raise taxes.[24] In the meantime, however, other major American cities were busily abandoning their reliance on private water suppliers. As early as 1798, for example, Philadelphia had pioneered the development of public water supply, followed by Cincinnati in 1817.[25] Finally, in 1833, a Water Commission was appointed by New York State to undertake the first systematic study of the city's water needs, deploying the latest advances in the geological and engineering sciences. This more rigorous approach coincided with a major cholera outbreak and enabled a consensus to emerge among New York's political and business elite that a source of water from outside the city had to be found. A number of different proposals were debated at the time, including the damming of the Hudson River, the tapping of the Passaic River in New Jersey, and the construction of a canal from the Bronx River into Manhattan. After much deliberation, technical opinion agreed on a plan to divert water from the Croton River forty miles to the north of the city using a closed masonry aqueduct.[26]

How can we account for the shift in technical and political opinion that occurred in the 1830s? A first development was the increased seriousness of water-related disease outbreaks in New York at a time when new evidence from its economic rival Philadelphia revealed that the construction of an improved water supply system had successfully reduced the incidence of disease.[27] A further issue was the growing status and professionalization of engineers, who were able to convince political and economic elites that a more technically ambitious solution to the city's water problems had to be found. The city's reluctance to invest in the project on the grounds of cost was countered by new evidence collected by the state-appointed water commissioners that at least 30,000 owners of building lots would be willing to subscribe to a new water service and easily cover its estimated construction cost of $5.4 million. Another critical factor to emerge in the 1830s was the growing extent of damage to property by fire, which threatened the city's burgeoning insurance industry. Fire damage was now costing the city more than any notional expenditure on a new water system, particularly in the wake of the "great fire" of 1835 in which 674 buildings were destroyed (figure 1.3).[28] Since directors of leading insurance companies were represented on both the state-appointed Water Commission and the city's Common Council, this must have intensified the sense of urgency to take action. There was also at this time increasingly frantic lobbying by chemical works, breweries, tanneries, distilleries, hotels, sugar refineries, and other parts of the fast-growing New York economy which relied on pure and reliable water supplies and would be ruined if the city authorities failed to take action.[29] The situation was merely exacerbated by the gathering pace of industrialization, urbanization, and the use of steam power, which accelerated the shift of industrial production to new urban centers. In the absence of a secure water supply, New York's boosters claimed, the city would lose out to its main rivals and future investment would be stymied.

In 1834 the State Legislature finally passed a law that gave the city the right to construct its first municipally owned waterworks. This proposal was then strongly approved in a citywide referendum the following year. In 1835 the city authorized an initial bond issue of over $2 million which would allow construction of the Croton Aqueduct to proceed. The first bond issue sold well in both the United States and Europe, despite the economic downturn of 1837–1843, and the city was easily able to complete the project to the satisfaction of its investors.[30] Predictably, almost all the opposition to the new water system came from uptown residents and property owners whose wells were not yet polluted, and their well-organized opposition succeeded in delaying construction for a further two years. Toward the source of the Croton River in Westchester County, residents "vigorously opposed the aqueduct, claiming it disfigured their fields and divided property."[31] For the historian Eugene Moehring, however, the successful completion of this project demonstrated how "the city had triumphed over a water shortage that had threatened its health, property, and prosperity. Despite

1.3 The great fire of 1835: lithograph published by H. R. Robinson (1836). In the absence of adequate water supplies, the city was under constant threat of fire.
Source: The Museum of the City of New York.

staggering costs and political opposition, authorities had met the crisis directly, setting an example for other towns. Problems had been legion—contract fraud, court suits, jurisdictional disputes, and pipe location, but New York overcame them."[32] The construction of the Croton Aqueduct marked a new era in North American urbanization. Had it not been built, it would have been impossible for New York City to retain its position as the largest and fastest-growing city in America.

1.2 Engineering the Technological Sublime

> Nothing is talked of or thought of in New York but Croton water; fountains, aqueducts, hydrants, and hose attract our attention and impede our progress through the streets. . . . Water! Water! is the universal note which is sounded through every part of the city, and infuses joy and exultation into the masses, even though they are out of spirits.
>
> —*Philip Hone*[33]

The completion of the Croton Aqueduct in 1842 was marked by the biggest public celebrations in New York City since American independence. The diarist Philip Hone recounts the five-mile-long procession as one of "perfect order and propriety," which he attributed to "the moral as well as the physical influence of water."[34] Since the writings of Vitruvius, the beauty and technical ingenuity of water-based architecture have been a recurring symbol of both prosperity and municipal independence (figure 1.4). The architectural historian Vittorio Gregotti, for example, describes how the aqueduct has, through history, created "a productive dialectics with the built fabric of the city" which has enabled "the unity of the *urbs* and the *civitas*":

> To supply water freely to a city is much more than guaranteeing a service: it represents, in an exemplary way, the collective effort to ensure the communal life of the settlement. It imposes its necessary

1.4 C. Bachman, lithograph of New York City (1849) showing the Croton water fountain in Union Square, which was constructed to celebrate the city's new water system.
Source: Collection of the New-York Historical Society.

> geometry and reconnects city and territory, geography and settlement. . . . Necessity, ingenuity and civic virtue seem to be represented in the aqueduct by an organic synthesis.[35]

The creation of New York's water system consolidated the emergence of a more sophisticated kind of urban society within which fragmentary and parochial perspectives were superseded by a more strategic urban vision. This new outlook was reinforced by impressive engineering feats in the service of a modern metropolis. The Croton Aqueduct incorporated the latest advances in French and British engineering and also added unprecedented features of its own. The elaborate new structure excelled the Roman aqueducts in the size of its cross section and also advanced on Assyrian and Roman structures by the use of the siphon: its unique features included the low 12-foot inverted siphons at the High Bridge and the one at the Manhattan Crossing that put the water under pressure.[36] The chief engineer for the project from 1836 was John B. Jervis, who had gained extensive experience from the construction of the Erie Canal (1817–1825). Jervis was one of a number transitional engineers who played a significant role in the modernization and professionalization of engineering science in the nineteenth century.[37]

Like the Erie Canal, the Croton Aqueduct also contributed toward a new kind of mediation between technology and nature in the American landscape. The economic growth of the city was increasingly tied to a regional urban ecology within which "wild nature" would be gradually displaced by an intensely reworked landscape. With the extension of the city's ecological frontier into upstate New York, the urbanization of nature and the naturalization of the urban could advance in tandem. The construction of aqueducts, dams, and reservoirs in upstate New York marked the evolution of a new kind of technological and cultural engagement with nature. "By the middle of the nineteenth century," argues the historian David Nye, "the American sublime was no longer a copy of European theory; it had begun to develop in ways appropriate to a democratic society in the throes of rapid industrialization and geographic expansion."[38] Crucial to this emerging aesthetic sensibility was the gradual supplanting of nature-based sublimity by an emphasis on the growing scale of human artifice.

The new water infrastructures presented an architectural vision quite different from the kind of romanticized native American landscapes depicted in the contemporary art of Thomas Cole, Asher Brown Durand, and other artists associated with the Hudson River School of painting. The latest accomplishments of the engineering sciences emerged as a field in which America could rival the Old World in its Promethean transformation of nature. The vibrant combination of modernity and Arcadia presented a symbolic concretization of the pastoral landscape in the service of republican ideals. Features of the Croton Aqueduct such as the High Bridge soon became recognizable icons in the New York landscape and markers on a path to a more clearly defined sense of national identity (figure 1.5).[39]

The construction of a new water infrastructure instituted a different kind of relationship between the city and nature. At one level the improved flow of water contributed toward a democratization of nature. Fountains and other architectural features symbolized the new urban bounty of fresh water brought from upstate sources; plentiful supplies of water helped to keep streets clean and contribute toward the creation of a more "hygienic" urbanism; while the gradual diffusion of plumbing technologies within the home lessened the daily burden of fetching water from standpipes and other sources. At first, some commentators were skeptical: the lawyer and diarist George Templeton Strong feared that Croton water would be full of "tadpoles and animalculae," to say nothing of "Hibernian vagabonds" relieving themselves into the aqueduct as they toiled on the completion of the project. But within a year, Strong took delight in the new bathroom fixtures installed by his father: "I've led rather an amphibious life for the last week," he wrote, "paddling in the bathing tub every night and constantly making new discoveries in the art and mystery of ablution. A real luxury, that bathing apparatus is."[40] Water gradually entered urban consciousness in a variety of ways, some public and some private, and in time the growing use of water would be seen as an indicator of modernity.

Despite the better access to water for many ordinary New Yorkers, the political impetus behind the construction of the new water system remained firmly economic in its motivation. The historian Joanne Goldman, for example,

1.5 Fayette B. Tower, lithograph depicting the Croton Aqueduct at the Harlem Bridge (1843). Source: The Museum of the City of New York.

suggests that the Croton Aqueduct marked a dramatic departure from the starkly unequal distribution of municipal services in the past, yet this observation oversimplifies the transformation of urban infrastructure.[41] The dramatic expansion in the scale and cost of public works in nineteenth-century New York was firmly grounded in an economic logic that found powerful political advocates. If impoverished Irish and German wards had received water before the wealthy residents of the Upper West Side, this was simply an anomalous outcome of the speed with which new pipes were constructed under the more densely populated parts of lower Manhattan.[42] The modernization of nineteenth-century cities in Europe and North America was not carried out in order to improve the conditions of the poor but to enhance the economic efficiency of urban space for capital investment. In this sense, the scale of new public works and the pace of technological change masked the persistence of social and political inequalities that would not be tackled in any systematic way until many decades later.[43] Advances in public health were an ambiguous by-product of the bourgeois rationalization of cities. Whatever the complex motivations that lay behind the development of elaborate public works projects, however, they did provide thousands of laboring jobs for native and immigrant workers whose votes were integral to the growth of machine politics with the widening of the political franchise. And even if the spread of new advances in public health was initially highly uneven, it did lend a powerful legitimacy to the development of new kinds of municipal governance freed from the *noblesse oblige* of the past.

So popular was Croton water that by 1850 New York could boast of the highest levels of per capita water consumption of any city in Europe or North America.[44] Earlier suspicions toward extensive water use were supplanted by a new enthusiasm for its diverse therapeutic and hygienic applications. Contemporary advocates of hydrotherapies such as Joel Shew saw the growing use of water as an indicator of the "general advancement of civilization" through American society.[45] Twentieth-century commentators such as Sigfried Giedion elaborated on this sense of dynamic optimism surrounding the spread of water through modern societies: "Words are too static," writes Giedion. "Only a moving picture could portray water's advance through the organism of the city, its leap to the higher levels, its distribution to the kitchen and ultimately to the bath."[46] Water use in American society grew steadily through the installation of new water technologies such as flush toilets and fixed washbasins, which came into general use from the 1850s onward (figure 1.6). Yet the spread of plumbing technologies can be attributed to changing fashions in health and architectural design rather than simply the greater availability of running water. The new popularity of water within the home is best conceived as a "private manifestation" of the impetus for greater personal space and new standards of hygiene.[47] As the historian Alain Corbin has pointed out in a French context, the growing use of water and the concomitant emphasis on cleanliness formed part of a complex pattern of cultural changes in the pre-Pasteur era associated with sharpening social and economic differentiations.[48] Water use became entangled in wider ideological discourses surrounding the promulgation of middle-class domesticity in the face of increased social polarization in the nineteenth-century city.[49] After all, the "water revolution" was initially largely restricted to the middle classes, with most working-class tenements lacking bathing facilities until legislative changes in the early twentieth century. Only in 1870, for example, did the city finally open two free public baths after decades of debate over personal hygiene (private bathing houses had already been in operation since the eighteenth century).[50]

As for human wastes, there was little consensus over the relative advantages of water closets and earth closets until increased water use in the late nineteenth century began to overwhelm the use of facilities unconnected to the sewer system.[51] The declining use of earth closets in the cities of Europe and North America marked a transition away from the circulatory preoccupations of the organic city and the desire to use human wastes as fertilizers. As increasing quantities of water entered these self-contained sewage systems for individual dwellings, the nitrogen content began to fall. "It is now conceded," wrote the New York sanitation pioneer George Waring in 1895, "that the very small amount of manure and the very large amount of water cannot be separated at a profit."[52] With the advent of modern plumbing systems, these earlier efforts to recycle human wastes

1.6 E. Jones, lithograph advertising new plumbing technologies (1845).
Source: The Museum of the City of New York.

were superseded by a new emphasis on large-scale technological systems to facilitate the flow of water through cities, which led ultimately to the development of sewage treatment works and other advanced methods of water purification and pollution control.[53] The dwindling use of privies, cesspools, and other ad hoc solutions to the problems of urban sanitation marks the decline of the "private city" where emphasis was placed on minimalist and fragmentary forms of municipal government. The widening political franchise saw new legislative efforts in 1879 and 1901 to improve the quality of tenement housing and ensure that the benefits of new plumbing technologies would be incorporated into building design. New patterns of water use became part of a wider transformation in living conditions which furthered the technical and cultural agenda of urban reformers and public health advocates as they sought to build centrally managed urban systems. For sanitary inspectors such as Robert Newman, the complexity of plumbing individual homes was a microcosm of the challenge to transform the circulatory dynamics of the entire city:

> No community and no city can preserve a wholesome condition without a supply of pure water; and an equally thorough purification from all refuse. To properly arrange this double circulation in a large house, is a matter of no trivial consideration; how much more, then, is skill, sagacity, and system, necessary for the sufficient supply and drainage of a district of an immense city like New York?[54]

By the early twentieth century the United States would be one of the best-plumbed nations in the world through the rapid diffusion of technologies such as pedestal sinks, enameled double-shell tubs, and siphonic-jet toilets.[55] An international survey of urban water consumption in the 1890s found that rates of water use in American cities were far higher than in those of continental Europe: only southern American cities such as New Orleans, with limited connections to urban water supply systems, exhibited lower rates of per capita water use than major European cities such as London, Paris, and Glasgow. We also find that New York's midcentury lead in per capita water consumption had been rapidly outstripped by fast-growing industrial cities such as Chicago, Pittsburgh, and Philadelphia.[56] Of course, the manufacturers of water-using technologies, plumbers, soap makers, and other industries that stood to benefit from an expanded use of water also played a role in fostering the spread of new consumption patterns: cleanliness was but one facet in the development of new cultures of retail shopping and advertising that would transform the lives of ordinary Americans.[57]

By the early 1870s a combination of droughts, low water pressure, and continuing urban growth, along with allegations of widespread graft and corruption, began to focus public attention on the need to rethink the scope and management of New York City's water supply system. As a result of the 1876 drought, for example, there was the first serious recognition that the repeated calls for water metering and repair of leaking pipes would not be sufficient to forestall serious future shortages. And by the 1880s supplies were so inadequate in the summer months that water could not be obtained from taps on the upper stories of tenement blocks. As a result of looming water shortages, a new phase of construction began for the Croton system in 1883. The New Croton Aqueduct (completed in 1890), the New Croton Dam (completed in 1907), and the Croton Falls Reservoir (completed in 1911) were the most elaborate water infrastructures ever constructed up to that time (figure 1.7). Had they not been undertaken, it might have proved catastrophic for the rapidly growing city: in 1895, for example, a severe drought had left most storage reservoirs almost empty.

By the early twentieth century the United States was the world leader in the building of dams for water supply, power, irrigation, and the control of navigation, yet as the historian Charles Weidner remarks, "most New Yorkers were too preoccupied with their city's growth to become unduly excited about their new aqueduct."[58] The New Croton Aqueduct was soon dismissed as simply one of the city's many achievements; it could hardly compare with the grandeur of other engineering projects such as John Augustus Roebling's design for the Brooklyn Bridge, which opened to international acclaim in 1883. The reconstruction of the "invisible city"—the upstate reservoirs, the underground pipe galleries, the valve chambers and other largely hidden or distant architectural

1.7 Croton Falls Reservoir, main dam under construction (1909).
Source: Collection of the New-York Historical Society.

features—could no longer capture the public imagination in the way they once had. "Municipal growth, whether slow or rapid," wrote the engineer James C. Bayles, "usually occurs by stages which are scarcely perceived, or at least scarcely realized, by citizens with whom it is a matter of daily experience."[59]

The completion of the Croton system did not solve the city's water problems. As early as the 1860s there had been fears for the safety of the city's new water system as a result of human and industrial wastes entering upstate streams and reservoirs. This marked an important advance in the epidemiological understanding of waterborne disease. In 1868, for example, Dr. Elisha Harris expressed concern over the "defilement" of the Croton system with sewage. Harris wrote widely on the subject and drew attention to the latest scientific debates in Europe and the need to disinfect drinking water by boiling.[60] A survey by the New York State Board of Health in 1884 revealed that villages, farmhouses, and mills in the Croton valley were draining their sewage directly into the river and its tributaries. As a result, the State Legislature granted the State Board of Health a series of new powers to control pollution in the city's watershed, but these proved relatively ineffective. Eventually, in 1893, new legislation was introduced that proposed to eliminate pollution by the acquisition of land along the streams of the Croton watershed. Despite contemporary descriptions of the water as turbid and strong-smelling, the city's newly established bacteriological laboratory assured the public in the early 1890s that the water presented no threat to public health.[61] But expert opinion remained divided between greater watershed protection and the early introduction of new filtration technologies. In 1894, for example, the engineer and entrepreneur Arnold Ruge lobbied the city authorities to introduce the latest French and Swiss filtration technologies on the pretext that the rich were simply buying themselves out of the problem of deteriorating water quality: "The rich in this city drink spring waters, imported from other States and even from Europe, but the masses of this City—the poorer classes—are compelled to drink unfiltered, dirty and even odoros Croton water."[62] In the event, a wide range of changes were instituted, and the construction of the more distant and higher-quality Catskill system was begun, which allowed the city to avoid any need to filter its water supplies until the dramatic reemergence of the water quality debate in the 1990s. During the first two decades of the twentieth century the emphasis on improving or protecting water quality was advanced further with a variety of initiatives. The most significant change was the use of water chlorination from 1910, which led to a sharp fall in recorded cases of typhoid and restored public confidence in the safety of the city's water.[63]

With the rising status of engineering, planning, and other new public service professions, urban management became increasingly influenced by a technical elite devoted to the rationalization of cities. In the wake of the Chicago exposition of 1893 and new developments in city planning pioneered in Sweden, Germany, and other industrialized countries, there was a strong convergence of international technical opinion around the need for more sophisticated and scientifically based modes of urban governance.[64] By the early decades of the twentieth century engineers like George T. Hammond were celebrating their role in the "wonderful growth of the urban." For Hammond, writing in 1916, there was no doubt that engineers must play a didactic role: "We are employed by the public, not only to do their work but also to lead them in technical municipal affairs. It is our duty and province to instruct as well as serve."[65] The rationalization of urban space became a kind of Taylorization transferred from the factory to the city: the creation of a scientifically organized world where Thorstein Veblen's dream of "engineers in power" might ultimately be realized.[66] It would be misleading, however, to conceive of the growing professionalization of technical opinion as constituting an undifferentiated perspective on the future of city management: important differences existed over rival technical and organizational solutions to the management of urban space.[67] Debates over technology and urban planning are best perceived as crisscrossing a complex web of evolving interests in the context of continuing elite domination of urban politics. Tensions existed between rival fractions of capital and also between different tiers of state authority as successive political machines vied for control over urban government.[68] The 1898 merger between the five boroughs expanded the scope and responsibilities of New York City's water supply system overnight, as the Croton system began to replace inferior alternatives such as the polluted wells of Queens, Staten Island, and the Bronx, as well as Brooklyn's surface-fed Ridgewood system based in Long Island.[69] From the 1890s onward a policy consensus gradually emerged that the Croton system could not sustain any major expansion in the future and that the only solution to the city's long-term water needs would be to use the more distant sources of the Catskill Mountains.

Urban growth instituted a brutal logic of its own which necessitated a transformation of the physical landscape over a vast area, introducing forms of strategic decision making beyond the scope of existing municipal government. An emerging political ecology of power linked the city to an ever greater swath of upstate land as part of a giant metabolic urban system. In order to begin this new phase of construction, the State Legislature created a powerful new structure in 1905 called the Board of Water Supply, which institutionalized the role of engineers in municipal government. This new body had immense powers: it could take over private land for water supply; it could work in relative autonomy from elected municipal government; and, following amendments to the state constitution, its projects were not limited in cost by the requirement to restrict bond issues to a specified percentage value of city real estate.[70] Foreign observers such as the British engineer Gilbert J. Fowler were amazed at the scale of this new undertaking:

> The world has been startled by the magnitude of your water schemes. Any European city would have regarded 114 gallons *per capita* as an extravagant allowance, yet this—which is equal to a daily supply of 500,000,000 gallons—is what is now obtained from the Croton works alone, but rather than curtail that supply, or do anything which might be interpreted to favour a limited use of water for public health purposes, the authorities determined to carry out a gigantic scheme to obtain another 500,000,000 gallons of water a day, this time from the Catskill mountains. The magnitude of

> the undertaking and the aggregate cost of bringing in 1,000,000,000 gallons per day will place New York water supply in a category by itself when a history of the world's great water works comes to be written.[71]

Before New York could commence an expansion of its water supply, however, it had to overcome what Charles Weidner describes as a "scandalous prelude" created by the attempt of the Ramapo Water Company to gain control over the water resources of the Catskill Mountains in anticipation of the city's future needs.[72] In 1895 the Ramapo Water Company had successfully lobbied the State Legislature to grant it land and water rights in the Catskill Mountains. As a result, New York City came very close to losing any control over the future design or cost of its water supply system. When the nature of the Ramapo Water Company's proposals became known, a bitter public response ensued led by City Comptroller Bird S. Coler.[73] In 1901 the State Legislature repealed the company's charter, thereby freeing the city from dependence on it for the future of its water supply. For some years, however, the company continued its attempts to overturn the decision of the State Legislature. As late as 1915, for example, the US Supreme Court dismissed a suit brought by the Ramapo Water Company that sought to prevent the construction of the Ashokan Reservoir, the first of the city's reservoirs to be constructed in the Catskill Mountains.

The first part of the Catskill system, constructed between 1907 and 1917, was far bigger than the Croton system. It includes the Ashokan, Kensico, Hill View, and Silver Lake reservoirs, as well as 126 miles of aqueduct (some 18 miles of which were bored through solid rock between 200 and 750 feet under the Harlem and East rivers and the streets of Manhattan).[74] Under the McClellan Act the city sought to acquire land without a repetition of the "rapacious proceedings" that characterized the use of land in the Croton watershed, yet bitter conflict ensued over the indirect loss of earnings from destroyed businesses and undervalued property acquisitions. The development of New York's water system led to the mass displacement and destruction of many settlements across the city's watershed. Upstate New York experienced its own water wars to rival that of the Hetch-Hetchy and Owens valleys in California.[75] In 1908, for example, Justice A. T. Clearwater publicly reprimanded the city for the betrayal of the people of Ulster County, whose land claims had been "scoffed and sneered at, derided and belittled." In the summer of 1913 a correspondent for the *Kingston Freeman* described the disappearance of the village of West Shokan to make way for the Ashokan Reservoir: "Very few buildings are left now to be burned. The trees are all cut down and the village is fading as a dream."[76]

The Ashokan Reservoir alone covers an area of 12.8 square miles, equal to the whole of Manhattan Island below 110th Street, with a capacity of over 130 billion gallons drawn from a mountainous catchment area of 257 square miles. The Catskill Aqueduct is twice as long as the greatest Roman aqueduct, being over twice the length of the two Croton aqueducts combined, and was designed by "a corps of engineers and experts unequalled in the history of engineering."[77] In 1917 Mayor George McClellan compared the construction of the Catskill Aqueduct to that of the Panama Canal. The comparison is apposite, considering the fact that the possibilities for large-scale urban reconstruction were facilitated by growing US economic and political hegemony in central and southern America at the time (see chapter 4). Urban infrastructure provided a reliable investment for new flows of capital, and continued urban growth was in turn related to vast social and environmental transformations that extended far beyond the US frontier.[78] In October 1917 the first Catskill water reached Manhattan and this event, like the completion of the Croton Aqueduct in 1842, was marked by a three-day celebration culminating in a rapturous reception in Central Park:

> At the Sheep Fold ceremony depicting American Indian tributes to "the good gift of water" a huge chorus of children and young women sang the National Anthem and the pageant was about two-thirds through when a downpour came. Strangely the rain fell just after the medicine men of the Indian Village and the priests of the ancient Orient, in compliance with the programme, had prayed for

> rain for benefit of the crops. The prayers had no sooner been uttered than the drops of water heralded the cloudburst. The children, laughing at the coincidence, scattered in all parts of the park.[79]

Almost as soon as the Catskill system had been completed in 1927, however, it became clear that an even bigger water source would be needed for the city. During the 1920s the Board of Water Supply carried out "endless investigations" of the more distant parts of the Delaware watershed.[80] In 1930, in anticipation of the city's water needs, the state of New Jersey sought an injunction in the US Supreme Court in order to prevent New York City from taking further water from the tributaries of the Delaware River (whose water supplied the needs of many New Jersey communities). In the event, the court found in the city's favor, with a decree granting the city permission to take 440 million gallons a day from the Delaware River as it passed through New York State.[81] In 1927 the city's Board of Estimate finally approved the Delaware project and authorized the issue of $64 million worth of city bonds, but the 1929 stock market crash delayed the start of construction until 1937.[82] The first phase of the Delaware system was eventually completed with a federal loan issued under the public works program of the New Deal, but the entire project was not completed until 1967 (figures 1.8 and 1.9). It is difficult to overestimate the significance of the New Deal for the modernization of the city's infrastructure: in 1940, for example, the chairman of the New York City Planning Commission, Rexford Tugwell, estimated that New Deal-funded capital improvements amounted to double what could have been achieved without federal assistance.[83] Thus the initiation of the final stage of the city's water system was enmeshed in a wider political and economic context for public works which increased the power and scope of large-scale infrastructure projects in urban policy making.

The period from the 1880s until the 1960s saw a continuous program of dam and reservoir construction for New York City. The idea that anything might restrict or impede the preeminence of New York as a world-class city proved unthinkable, and engineers were ready to provide ever more elaborate means to slake the "thirsty metropolis."[84] It would be misleading, however, to argue that

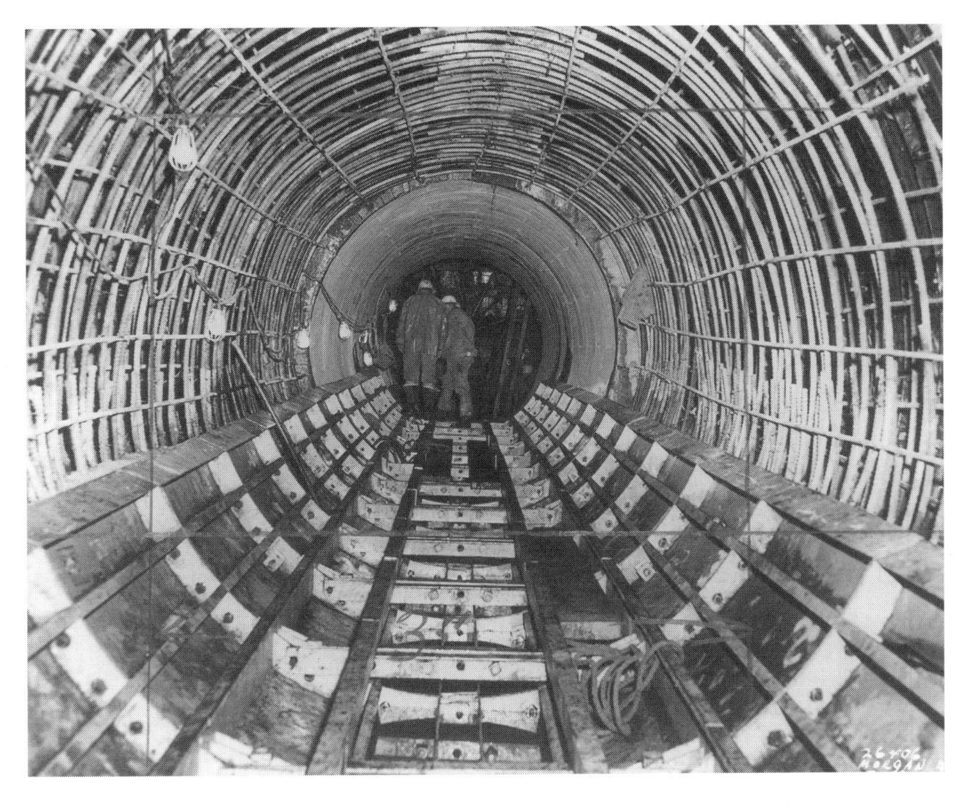

1.8 The Morgan Avenue Interceptor for the Delaware system under construction in 1958. Source: Collection of the New-York Historical Society.

1.9 Water supply engineers study the West Delaware Tunnel with a New York University computer, circa 1960.
Source: Collection of the New-York Historical Society.

engineers and city planners uniformly supported the logic of ever greater water use: there was a parallel ascetic discourse of disdain for the profligate use of water in the post-World War II era. The city engineer Edward J. Clark, for example, publicly derided the wastage of water by lawn sprinklers and children playing with fire hydrants.[85] The serious drought of 1949–1950 not only reinforced the urgency of the need to complete the Delaware project but also heralded intense water conservation efforts for the first time. By the early 1960s a further series of droughts necessitated the pumping of extra water from the Hudson River, and serious debate ensued over the mooted construction of nuclear-powered desalination plants as the only viable long-term water strategy for the city.[86] In the summer of 1965 the engineer Abel Wolman made a pointed contrast between the apparently drought-induced plight of New York and the increasingly sophisticated water infrastructures of semiarid southern California. Wolman railed against "delayed action and failures of management" which had led to the absurdity that while "New Yorkers were watching their emptying reservoirs and hoping for rain, Californians were busy building an aqueduct that would carry water some 440 miles."[87] For Wolman, water shortages in the modern era were the outcome of political vacillation rather than climatic perturbation.

The factors that determine the long-term viability of cities and regions rest ultimately not with natural limits, which are in any case largely culturally and technically determined, but with the strategic significance of places within a wider set of social and economic dynamics. And it is precisely this interaction between regional urban systems and the changing dynamics of capitalist urbanization that was to emerge as the key dilemma for water policy making in the 1970s. The earlier symbiosis between water scarcity and the power of engineering within urban management had become trapped within an anachronistic framework unable to grapple with newly emerging realities. We now find a shift away from preoccupations with adequate supplies toward newly emerging concerns with the safety of drinking water and the maintenance of physical infrastructure. The sense of technological and administrative omnipotence that pervaded the completion of the vast Catskill and Delaware systems had begun to fade.

1.3 Urban Decay and the Hidden City

> Space as threat, as harbinger of the unseen, operates as medical and psychical metaphor for all the possible erosions of bourgeois bodily and social well-being.
>
> —*Anthony Vidler*[88]

The creation, since the nineteenth century, of large-scale technological systems for the provision of drinking water, irrigation, and power is a pivotal element in the modern impulse to rationalize nature. The integration of the hydrological cycle into the circulatory dynamics of the modern city produced not only a material transformation in the built environment but also a symbolic landscape of political and cultural power.[89] By the late 1950s it appeared that the New York metropolitan region had "developed a kind of supremacy, in politics, in economics, and possibly even in cultural activities."[90] New urban infrastructures were an integral dimension to this hypermodernization of urban form within which speed, mobility, and economic efficiency took precedence over alternative urban visions developed around more vernacular or community-based modes of urban living. From the late 1960s onward, however, America's celebration of technical progress had begun to shift decisively away from vast civil engineering structures to new advances in electronics, computing, and bioengineering.[91] The decline of the Fordist city, with its associated regional clusters of manufacturing industries, was accompanied by a reemergence of latent contradictions within the modernization of urban space. Fears of crime and social disorder associated with the chaotic and anarchic spaces of nineteenth-century urbanism began to resurface in response to the urban crises of the late 1960s and 1970s that swept through American cities. New York began to exhibit a kind of spectral urbanism in which the ghostly traces of former cycles of investment coexisted with new cycles of investment in the built environment. A crumbling mass of infrastructure from the past encircled the new architectural symbols of the city's global preeminence. The fountains of the Kensico Dam, to the north of the city in Westchester County, for example, lay dry and neglected as a forgotten piece of the city's past, the cracked and discolored stonework covered by a patina of mold like the ruins of an ancient civilization.[92] And by 1982 the city's water commissioner, Francis X. McArdle, was warning of "decline and disintegration" for the city's water system.[93]

The shifting relationship between culture and technology in New York can be traced to identifiable changes in patterns of infrastructure investment that stem from the fluctuating economic fortunes of the city within an increasingly globalized system of capital mobility and economic production. A fundamental dynamic here is the emerging disjuncture between the mobility of capital and the extent of fixed capital represented by past investment in the built environment. In the 1830s we saw how these two dimensions were successfully combined in order to secure a degree of synergy between urban form and economic prosperity through the early development of capital markets and new administrative structures. Subsequent economic downturns in the 1850s, 1870s, and 1920s all failed to seriously undermine this relationship but led to new configurations in the role of the state for infrastructure provision in order to sustain continued urban growth. In the mid-1970s New York again faced an economic downturn. The city faced a severe fiscal crisis in 1975 derived from the cumulative impact of rapid depopulation and deindustrialization since the 1960s, which had led to a shrinking local tax base at the same time as there were demands for higher social expenditure to mitigate the social and economic impact of urban decline.[94]

The late 1960s and early 1970s saw the disintegration of any consensus over the aim and method of urban planning. This transitional period was marked by the declining power and prestige of technical and professional elites in urban government. A growing body of literature explored the hiatus between liberal and Marxian interpretations of persistent social and economic inequalities associated with the process of capitalist urbanization.[95] The perceived urban crisis of the late 1960s and 1970s shattered preconceived approaches to the management of urban space inherited from the Progressive era. The work of David Harvey, for example, provided important insights into the relationship between the partial recomposition of urban form and the spatial and sectoral switching of capital investment between different elements in the urban environment.[96] The political and economic turmoil facing European and North American cities undermined existing conceptions of the linkages between urban form, civil society, and the technical rationalization of urban space. Under post-Fordist urbanism the secondary circuit of capital investment, which includes the built environment, had become more sharply differentiated in both spatial and sectoral terms. In the case of New York these trends have been significant in two main ways: first, there has been an ongoing crisis in the maintenance of the physical fabric of the city; and second, a decisive shift has occurred in the way capital investment is funded, with a growing dependence on socially regressive sources of revenue such as user charges. By the 1980s, however, the radical impulse behind the critique of liberal planning discourses had been supplanted by a new emphasis on the power of private capital to shape urban space. The role of planners, engineers, and public policy advocates had been increasingly eclipsed by bankers, lawyers, and bond underwriters determined to redirect urban governance toward the needs of business. The relative significance of the public city and the private city had been altered by changing patterns of investment in the built environment that have consistently favored the promotion of new opportunities for capital accumulation in the place of the now much-maligned public agenda of the New Deal era. Yet as we shall see in subsequent chapters, the public realm fostered under the New Deal was in many respects a trajectory of private aspirations that rested on sharp contradictions in the urban experience. The story of New York's water supply represents one element in this changing urban dynamic, as the existing rationale for large-scale urban policy making became progressively disengaged from the wider dynamics of urban and regional change.

The altered political and economic circumstances for urban policy making in the 1970s were to have a profound impact on urban infrastructure. In the early 1970s, in the years preceding the 1975 fiscal crisis, we find higher levels of investment than in any subsequent period, marked by a predominance of federal and New York State sources of funding (figure 1.10). The 1975 fiscal crisis brought about a virtual collapse in capital investment: the period between 1975 and 1990 effectively represented a disinvestment in physical infrastructure, with levels of expenditure falling below that required to maintain the value of the city's capital

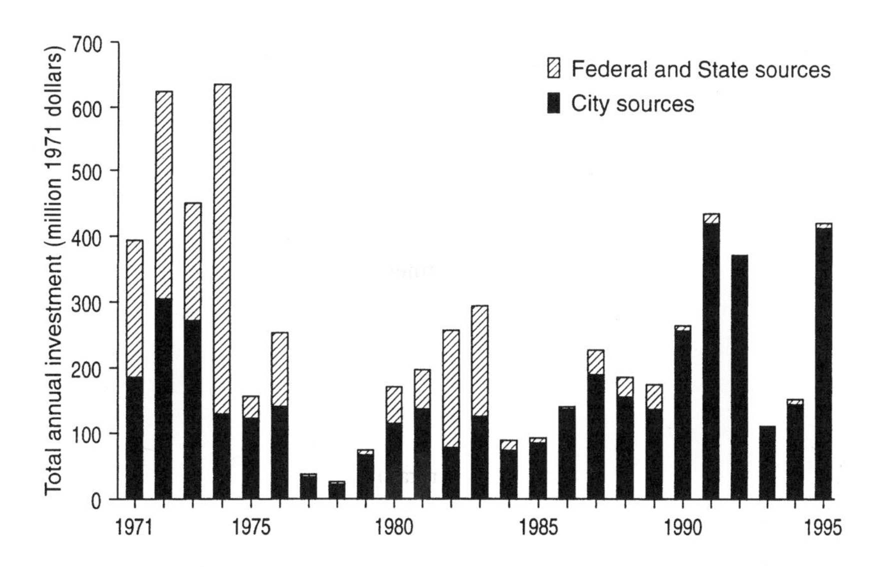

1.10 Changing levels and sources of capital investment in the New York City water and sewer system, 1971–1995.
Source: New York City Record, New York City Office of the Comptroller and New York City Office of Management and Budget.

stock. In the wake of the 1975 fiscal crisis New York City found itself shut out of the municipal bond market. As a consequence, by the late 1970s the city was completely reliant on grants from the federal government and New York State for its capital budget.[97] The "hidden city," represented by the vast networks of pipes, tunnels, bridges, and other basic elements in an urban infrastructural palimpsest, bore the brunt of a reallocation of resources away from long-term capital investment in urban infrastructure.

In the early 1980s there was a modest recovery aided by a new input of federal funding, but by the mid-1980s the level of capital investment had again begun to falter: between 1977 and 1985 the overall contribution of federal expenditure to the city's budget fell sharply from 19 percent to 9 percent.[98] Since the end of the 1980s there has been a major expansion in capital investment from city sources, with renewed levels of investment in the 1990s derived from a fundamental reconfiguration in the arrangements for capital funding through an increased reliance on municipal bond markets. Yet despite the increased level of capital investment, the city has still not reached the level of investment of the pre-fiscal crisis era. Furthermore, current levels of investment are not adequate to prevent a further deterioration in the city's infrastructure. The median age of water mains and sewers continues to rise, resulting in a gradual yet inexorable increase in numbers of water main breaks. Over half of the city's water mains consist of unlined cast iron laid before 1930, whereas only 9 percent are composed of more flexible ductile iron laid after 1970, which is less likely to crack under stress. By the year 2020 more than 35 percent of the city's water mains will be over 100 years old, and the city will face severe difficulty in maintaining its infrastructure without punitive increases in water and sewer fees.[99] The consequences of dilapidated infrastructure are that nearly 5 percent of the city's water supply is estimated to be lost through leaks, undermining efforts to conserve water, and the city is repeatedly faced with expensive pipe failures, in some instances involving substantial flooding of the subway system and extensive road collapses costing many millions of dollars to repair.[100]

With the sharp decline in federal and New York State funding for water supply, the city has become increasingly reliant on just two sources of money:

user charges for water and sewer services, and the issue of municipal bonds specifically for the water and sewer system. The increasing reliance on revenue financing from water and sewer fees to support capital investment mirrors similar developments in a number of larger US cities including Boston, Chicago, Philadelphia, and Detroit. Reluctance to raise property taxes since the 1970s has contributed to a rise in water and sewer fees of nearly 200 percent between 1987 and 1994.[101] By the late 1990s nearly a quarter of the city's water bills were in arrears, leading to the enactment of new regulations giving the city the right to shut off water supply to individual buildings: a threat that now hangs over much of the city's poorest multioccupancy low-income housing.[102] The background to this change can be traced to new funding arrangements set up under Mayor Edward I. Koch in 1985 with the creation of two quasi-autonomous city agencies, the New York City Municipal Water Finance Authority and the New York City Water Board, in order to relieve some pressure on the city's capital budget. The Municipal Water Finance Authority borrows money for capital projects by selling bonds backed by water and sewer fees. The Water Board then sets water and sewer rates at whatever level is necessary to pay back the bonds and run the water supply system.[103] In addition to political pressures to hold down property taxes, a number of other factors have contributed to the sharp rise in water and sewer fees since the late 1980s: the increasing costs of running the water supply system; a continued decline in federal and New York State sources of funding (especially since the election of the second Reagan administration in 1984); the unwillingness of voters to approve additional construction bond issues (epitomized by the heavily defeated proposal for an environmental bond issue in 1989); and increasing debt service on newly issued revenue bonds. In poorer parts of the city, rising water charges during the 1990s have begun to threaten the economic viability of low-cost housing in the multioccupancy private rented sector (a threat that has also been recognized in other American cities such as Denver, Detroit, and Los Angeles).[104] What is certain is that the era of cheap unlimited water use is over. In the wake of further droughts in the 1980s, and the gradual introduction of water metering since 1988, the relationship between the city and its water supply system has fundamentally altered. The prospect of a new phase of expansion in the city's water infrastructure to tap still more remote sources in upstate New York has been tempered by a new kind of fiscal and environmental austerity in water use. Available evidence suggests that water use has indeed stabilized since the 1980s after decades of rapid growth (figure 1.11).[105]

In 1993 the city's newly elected Republican mayor, Rudolph W. Giuliani, marked a decisive move toward greater fiscal austerity and a shift toward greater reliance on the private sector for the provision of municipal services. Within two years Giuliani announced his intention to sell off the entire water system to the New York City Water Board, which currently leases the system from the city. Giuliani argued that this transaction would allow the city to do a number of things: pay off old bonds (to allow new borrowing); write off unrecoverable taxes; and release $1 billion over the next four years for urgent capital projects, enabling the city to exceed its current borrowing limit. The Water Board would fund the sale by issuing $2.5 billion in new water and sewer bonds with a debt service of some $2 billion, equivalent to around $65 million a year over a 30-year period. Giuliani's proposed sale of the water system led to fierce competition among Wall Street bond underwriters to win the $1.5 million in fees for managing the sale of the city's water assets.[106] However, this attempt to find a short-term solution to the city's budget deficit was blocked by City Comptroller Alan Hevesi, thereby exposing deep divisions in city water policy. Comptroller Hevesi castigated the proposed sale of valuable city assets in order to reduce the current budget deficit as a "fiscal gimmick" with deleterious long-term consequences for the city.[107] Hevesi argued that the loss of the water and sewer system would accelerate the erosion of the city's control of its watershed, making it easier for the state governor to change the composition of the city's Water Board in favor of upstate development interests (the seven members are currently appointed by the city's mayor). A succession of legal battles upheld these concerns and ruled against Giuliani, in a significant reverse for the neoliberal impetus toward the privatization of public services in American cities.[108] As a consequence of these political struggles over the control of public assets, the New York water system has emerged as a significant bulwark for the continuing importance of democratic accountability and regional coordination in the delivery of public services.

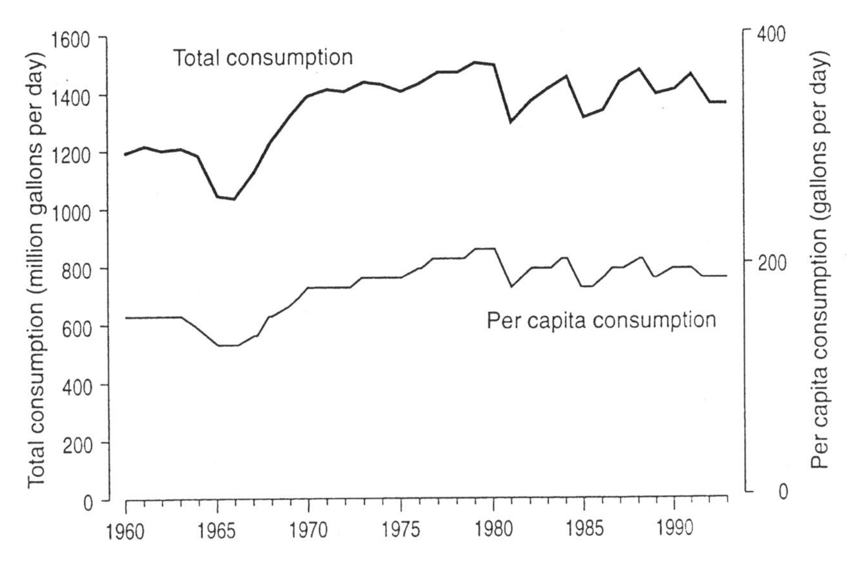

1.11 Changing levels of citywide and per capita water consumption in New York, 1960–1993.
Source: New York City Department of City Planning, with additional data adapted from R. Cropf, "Water Resources," in C. Brecher and R. D. Horton, eds., *Setting Municipal Priorities 1990* (London and New York: New York University Press, 1989), pp. 173–197.

1.4 Paranoid Urbanism

> The disconnection from our places of service is furthered by growing doubts about the sustainability of America's technological achievements. Our water, which we once saluted for its unrivalled purity and propitiated with splendid water towers and graceful dams, now seems to contain the specter of lurking microbes and toxins.
>
> —*Thomas H. Garver*[109]

The maintenance of physical infrastructure represents only one side to the city's water crisis. The other dimension is provided by declining public confidence in the safety of the drinking water: a situation unprecedented since the typhoid scares of the prechlorination era in the early twentieth century. By the early 1990s the safety of New York water had become the most contentious environmental issue in the city. In the early 1960s, by contrast, commentators on the extension of the Delaware system, then under construction, had actually suggested that the quality of water from the new Cannonsville Reservoir was "uneconomically" good: it was argued that the city should have looked to cheaper but inferior sources such as filtered water from the Hudson River.[110] The most distinctive feature of New York's water supply system over the last century has been the extraordinarily high quality of its unfiltered mountain water. In November 1991, for example, a *New York Times* article eulogized the city's water as a "great municipal blessing" and proudly noted that Fortnum and Mason's tea department in London would use only the purest water available: New York tap water.[111] A few days later, however, an irate letter to the editor of the *New York Times* from Assemblyman John Ravitz sought to draw public attention to dangerously low reservoir levels and a significant decline in the quality of the older Croton water system.[112] When the water supply system was originally constructed the city's watershed was remote and sparsely populated, but upstate development pressures have now thrown the long-term security of the city's water system into doubt. Water quality in the Croton system nearest to the city, which supplies around 10 percent of the city's water, has been steadily deteriorating since the early 1960s.[113] The concentration of the indicator pollutant chloride has risen from 8.8 mg/l in 1960 to 33.2 mg/l in 1990, an increase of 265 percent, in comparison with average concentrations of 8.5 mg/l in 1990 in the more distant and sparsely populated Catskill-Delaware system (figure 1.12).[114] Another key indicator of declining water quality is the presence of the fecal coliform bacterium *Escherichia coli,* which was detected in the summer of 1993 in the Chelsea and Lower East Side districts of Manhattan served by part of the Croton system. In the ensuing public health alert the city advised households in affected areas of the city to boil their water, which was accompanied by a sharp rise in sales of bottled water across the city as a whole (figure 1.13).[115]

Bottled water consumption in New York City has grown rapidly in response to skillful marketing of mineral waters as part of the rise of "health consumerism" coupled with increased anxiety over water quality in the 1990s.[116] But these trends are not explicable simply in terms of elite marketing: water quality has entered the wider arena of risk and uncertainty in public policy making, in consequence of an erosion of trust between citizens and the executive authorities of the state.[117] The very idea of safe public water supply has become a fantastical notion in the popular imagination. Sales of bottled water across the US have tripled during the 1990s despite evidence that public supplies are often of superior quality. Indeed, New York saw one of the fastest-growing markets for bottled water *before* the public health alerts of the 1990s.[118] We can identify a connection here in which the development pressures in the city's watershed lead to declining water quality while the polarization of incomes and lifestyles contribute to the increased consumption of elite consumer goods. A degraded public water supply system now operates in combination with increasing access to private sources of drinking water by the better-off, in a dangerous reversion to nineteenth-century patterns of service provision. The increasing use of bottled water by the rich also presents a bizarre inversion of the contemporary picture in developing countries, where the poor rely on expensive water vendors whereas the middle classes are connected to cheaper sources of piped water.

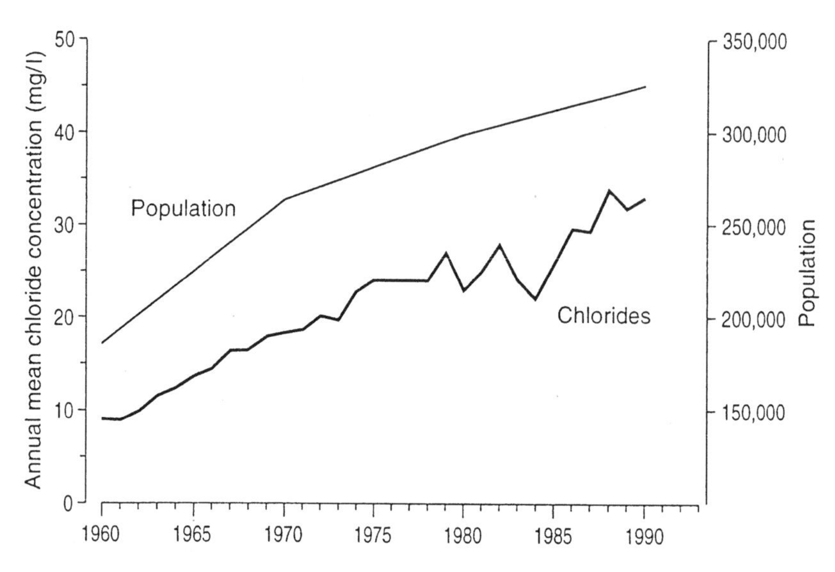

1.12 Changes in mean annual chloride concentration and population in the Croton watershed, 1960–1990.
Source: New York City Department of Environmental Protection.

1.13 The Giuliani administration has been periodically overwhelmed by the city's water crisis (*Village Voice,* 11 July 1995).
Source: Courtesy of Steve Brodner.

The most immediate problem facing New York's water quality has been quite specific: it is only in the smaller Croton system that a measurable deterioration in water quality has actually presented any threat to public health. Nine of the Croton system's 12 reservoirs are now overloaded with phosphates from inadequate sewer systems, and since 1990 the entire Croton system has been repeatedly closed due to summer algal growth.[119] Nonepidemic strains of cholera, as well as *Giardia* and *Cryptosporidium,* have also been detected in the city's reservoirs. Yet the picture of declining water quality is not as straightforward as it may appear: the increasing sophistication of environmental monitoring over the last twenty years has revealed the extent of previously undetected and largely unknown threats to public health such as new chemical compounds, *Cryptosporidium,* and other pathogenic organisms at the limit of detection.[120] The threat of *Cryptosporidium* raises three particularly perplexing issues for environmental regulation: the health uncertainties form part of a wider sense of public unease with the regulatory capacities of modern government; the risk lies at the edge of scientific knowledge (the organism was not recognized as a cause of human disease before the mid-1970s); and the contamination cannot be simply prevented by technological means through the filtering of public water supplies (as the Milwaukee outbreak of 1994 revealed).[121] The complexity of environmental regulation has also been intensified by the proliferation of so-called non-point sources of pollution, making individual polluters much harder to identify.

The background to the current escalation in political conflict over water quality can be traced to a tightening of federal water quality standards since 1989. As a result of these new standards the federal Environmental Protection Agency issued the city a deadline to prevent further deterioration in water quality for the Catskill-Delaware system by 2002 or be forced to undertake a massive program of filtration, at a cost of some $8 billion for the entire water supply.[122] Major engineering companies have been lobbying for this lucrative contract, which marks, for the private sector, a significant extension to potential investment opportunities in municipal service provision.[123] The new federal demands for technological improvements to water infrastructure have contributed to a market-led process of "ecological modernization" in which increasingly large shares of the city's capital budget are devoted to meeting ever higher environmental standards. The combination of higher charges with the extension of water metering to the poorest parts of the city is set to exacerbate the socially regressive impact of new patterns of capital investment, and even environmentalists have balked at the scale of this expenditure which threatens to undermine public support for environmental quality.[124]

The dilemma facing the city is whether filtration can be avoided for the Catskill-Delaware system without undermining regional development for some of the poorest communities in upstate New York.[125] As for the smaller Croton watershed to the east of the Hudson River, the political prospects of regulating what is now an affluent commuter belt have steadily deteriorated, with new pro-development groups emerging such as the Alliance for Watershed and Water Development that seek an end to current watershed regulations and promote filtration for the city's water.[126] Putnam County, located just 30 miles north of the city in the heart of the Croton watershed, is now the fastest-growing county in the state, with a stream of new developments under way for luxury homes, golf courses, offices, hotels, and shopping malls.[127] A number of these pro-development groups have been revealed to be significant contributors to the 1994 election campaign fund of Governor George Pataki, who has emerged as the most powerful political player in the drafting of new regulations for development in the city's watershed.[128] Although the battle to protect water quality in the Croton watershed has now almost certainly been lost, the future of the other 90 percent of the city's water system remains open to question. The water quality chemist Gerald Iwan, taking a historical view of developments, argues that the city faces a stark choice between "a future of complete dependence on treatment technology and associated economic and technical responsibilities of unimaginable magnitude" or a policy of enhanced watershed protection in order to maintain "cost-effective, high-quality drinking water without the complexities of superfluous treatment technology and source quality degradation."[129] The future of the city's water supply presents a microcosm of the whole gamut of challenges facing the future of environmental regulation given the declining significance of municipal governance in urban planning. The choice among different options remains delicately balanced, and as the city's power to shape developments has waned a new phase in the political ecology of urban water supply has begun to emerge.

The 1990s have seen an intensification of efforts by the city to protect its water supply, but the logistical problems are formidable. The watershed police—a designated branch of the New York City Police Department—were originally set up to protect upstate communities from workers involved in the construction of the city's water system and to prevent damage to city-owned property. Only during wartime, in response to fears of sabotage, have there ever been extensive efforts to guard the whole of the city's upstate network of aqueducts, dams, and reservoirs. For much of the postwar period, only a handful of police officers have been assigned to protect 2,000 square miles of watershed from many thousands of potential sources of pollution (figure 1.14).[130] There is a long-standing sense of regulatory ineffectiveness: the thirty-year period from 1960 to 1990 saw no prosecutions of water polluters despite evidence of declining water quality.[131] The lack of effective regulation in the city's watershed can be attributed to an ethos of regulatory laissez-faire shared by city-based engineers and upstate development interests.[132] Studies of environmental regulation have often revealed how technological solutions are favored by business interests and government agencies alike because they avoid reliance on politically contentious attempts to control individual polluters. And it is this question of the appropriate extent of urban authority over rural land use that is now at the heart of the regulatory dilemma facing the future of the city's water supply system.

The year 1990 saw the first efforts to revise the city's watershed rules since 1953. These new watershed protection rules address more recent threats such as pesticides, herbicides, and other toxins absent from the earlier regulations. However, the new watershed protection rules have been greatly weakened under political pressure from upstate interests and the Republican-controlled New York State Senate in Albany. A particular barrier to effective regulation of water quality is that the city owns less than four percent of the land in its watershed, far lower than the rate for other US cities such as Boston and Portland, Oregon, for example, which also rely on unfiltered water supplies.[133] The 1993 proposal by the city

1.14 Property of New York City: the Neversink Reservoir in 1995.
Source: Photograph taken by Matthew Gandy.

to buy 80,000 more acres of watershed land has run into repeated political and fiscal obstacles. The period since 1993 has seen a stark polarization between city and upstate interests, with the perceived regulatory interference in the watershed being met by a growing mobilization of property rights activists (with links to well-established groups in the Adirondack Mountains further north) and increasingly sophisticated antiregulation lobbies such as the Coalition of Watershed Towns. The possibilities for environmental regulation have been undermined by a deep-rooted ideological and cultural divide between urban environmental activists and rural upstate communities resenting outside interference in land use planning. Since the early 1990s the regulatory role of city and state authorities has come under sustained political attack from upstate communities and development interests in the watershed who demand less regulation, and also from the increasingly well-organized and vociferous city-based environmental groups who demand greater regulation and control of the city's watershed.

Under the watershed protection plan of 1997, the city has been forced to finance extensive economic development programs for the watershed communities over a period of fifteen years in exchange for greater cooperation in the prevention of water pollution. A different pattern of environmental regulation has emerged with the creation of novel institutional structures such as the Catskill Watershed Corporation and the Watershed Protection and Partnership Council, representing both city and upstate interests. In the place of a relatively centralized, ossified, and nonparticipatory regulatory system, the watershed is now overseen by a complex and dynamic jigsaw puzzle of different interest groups ranging from upstate lumber companies to city-based ecologists.[134] Yet behind this apparent opening out of environmental policy making, a more fundamental shift has occurred: the city has found its power diminished in relation to a broad coalition of forces ranging from agricultural interests to speculative real estate with a common interest in relaxing land use controls across the city's watershed. The diverse fractions of capital represented in the Watershed Protection and Partnership Council have successfully coalesced around a regional antiregulatory agenda capable of dominating political debate over the future of the city's water supply. New patterns of governance at a regional level have led to a historical shift in power away from the city-based traditions rooted in the municipal managerialist approaches of the past, with the regional dimensions to regulatory planning now extensively challenged by the upsurge of grassroots property rights activism. A variety of piecemeal and experimental interventions have emerged from the disintegration of a relatively stable mode of environmental regulation that had dominated regional water resources management since the nineteenth century. By 1999 there were already signs that the new watershed agreement had come under strain, with a further relaxation of controls on the use of crucial wetlands that act as natural buffers to minimize the contamination of streams and rivers. Powerful upstate real estate developers had also begun to mount a series of legal challenges to the new watershed protection rules using nineteenth-century state property laws. By the spring of 2000 the city had managed to acquire just 17 acres out of the 1,000 acres identified around the Kensico Reservoir in the heart of the fast-growing White Plains region as crucial for the protection of the water supply en route to the city.[135]

At the beginning of the twenty-first century the city is faced with an intensifying conflict between the short-term profitability of capital speculation and the long-term ecological viability of its watershed. The protracted negotiations between the city and its rural watershed in the context of global forces affecting the fiscal and political autonomy of the city reflect a new complexity in the political dynamics of urban water supply. It is now widely acknowledged that technocratic approaches to urban management, which reached their zenith in the late 1950s and 1960s, could never satisfactorily handle the complexity of democratic public participation. Yet the emergence of new institutional structures has become dangerously dislocated from the core decision-making processes that shape the development of urban space. In effect, wider public participation in decision making has been chimerical because legislative and regulatory agendas continue to shore up the regional needs of political and economic elites, as reflected in the growing power of upstate development interests over weakened urban government.

1.5 Hydrological Transformations

The water supply of New York has passed through a series of transformations since the early seventeenth century. The first period, lasting from the founding of the original settlement in 1626 until 1658, was marked by a reliance on natural water sources and private wells. A second phase, from 1658 until 1774, saw an expanding network of public wells within a context of steadily declining water quality. A third interval, from 1774 to 1830, was dominated by a series of ill-fated private interventions including the role of the infamous Manhattan Company. This chaotic urban scene was characterized by repeated outbreaks of disease, uncontrollable fires, and escalating economic disruption. A fourth phase, from 1837 to 1911, saw the construction and expansion of the Croton system as the city's first comprehensive public water supply. The modernization of the city's water system was marked by a series of advances spanning the bacteriological, technical, and administrative dimensions to water resources management, which mirrored developments elsewhere in Europe and North America at this time. A fifth period, between 1907 and 1967, marked the completion of the Catskill-Delaware system and an expanded role for municipal government in the management of regional water resources. This was the zenith of the technical management of urban space, with maximum power and autonomy for government agencies reached under the New Deal era. The most recent phase, extending from the late 1960s until the present time, has been characterized by a series of complex challenges to existing patterns of water provision. Regional economic change and new patterns of sociospatial restructuring have contributed to the emergence of a series of major policy dilemmas in the fields of capital investment and water quality.

The period between the completion of the Croton Aqueduct in 1842 and the city's fiscal crisis of 1975 marks a phase of remarkable stability in the history of New York's water supply. The nineteenth century saw a decisive shift from private to public water provision in order to allow new levels of efficiency and coordination. A series of tensions were played out not only between public and private interests but also among disparate bodies of technical expertise and rival political machines. During the twentieth century some of these disputes were resolved with a move toward the greater consolidation of fragmentary interventions to form powerful regional systems of management and control. This implied a partial waning of local democratic input, as technical elites emerged to design and operate vast public works systems. The creation of semiautonomous government structures fiscally and politically insulated from local electorates marks a smaller-scale precursor to the powerful regionally based federal agencies of the New Deal such as the Tennessee Valley Authority. The dams, reservoirs, and other large-scale infrastructure projects of the New Deal era have been widely interpreted as the epitome of American modernism. The combination of a utilitarian aesthetic in the International Style with a functional commitment to the rationalization of regional water resources became a symbol of a new kind of public landscape. These "democratic pyramids," to use Lewis Mumford's phrase, represent a unique conjunction of technology, nature, and public policy making, but their physical longevity belies the fragility of the cultural and political circumstances that facilitated their construction.[136] With the fading of the New Deal ethos in the 1970s, a new set of political, economic, and cultural developments began to shape the evolution of regional water policy. Recent changes are distinctive in a number of respects: the emergence of new sources of environmental risk such as cryptosporidiosis; the development of greater degrees of public skepticism toward technical and scientific expertise; the weakening of city power in relation to regional political developments; and above all, the intensity of the neoliberal challenge to the fiscal autonomy and ideological legitimacy of an effectively regulated and adequately funded public water system.

For over 140 years New York City successfully provided cheap, plentiful, and high-quality water to its citizens, on the basis of a settled relationship between water technologies and the "democratic urban landscape"; yet this historic achievement is now thrown into doubt by a series of political and economic developments beyond the reach of any regulatory or democratic structures yet devised. Some urban scholars have argued that social and economic developments since the 1970s have lessened any technical link between capital and urban form: the connection between urban morphology and economic function has become weakened.[137] In the case of water supply, however, this claim is problematic because of the continuing functional dimensions to urban space. The ongoing construction of the city's third water tunnel, for example, suggests that the material determinants of urban form may conform to a deeper logic than the more ephemeral political and cultural shifts surrounding the construction and design of real estate and other speculative elements in the built environment. Still, even if the technical dimensions to the design of urban space retain a high degree of continuity, the pressures to transfer public assets into the private sector have become immense.

During the 1990s the privatization of urban water systems gathered global momentum. In 1997, for example, the *Financial Times* proclaimed, "Water is the last frontier in privatisation around the world."[138] The sale of public water systems not only flows from the fiscal weakness of municipal authorities worldwide but has also been pushed by national governments in order to bolster foreign currency reserves and find favor with international financial institutions. The global marketization of water has not been without high-profile protests, as grassroots campaigns in Argentina (Tucumán province), Bolivia (La Paz and El Alto), Manila, and Barcelona attest.[139] While the New York case did not lead to mass protests, there is little doubt that the city's water supply has been politicized to a greater degree than at any time since the failure of the Manhattan Company in the early nineteenth century. What we are seeing in New York is a protracted process of reshaping the role of the municipal government in urban water supply. In effect, a hollowing out of government arising from a combination of fiscal and ideological pressures is leading to a polarization in the public policy debate between demands for water quality protection, advanced principally by urban environmentalists, and a coalition of antiregulation upstate interests, whose rhetoric is rooted in a legacy of land use conflict in the city's watershed.[140] The future form of environmental management is emerging as a politically contested reconfiguration of public policy, in this instance centered on a redefinition of the administrative powers of city government. At the heart of the debate over environmental management in the city lies a tension between market-led development pressures and the administrative jurisdiction of municipal authorities. The blocked sale of the city's water system in 1997 suggests that a stable new configuration of power between capital and municipal governance has yet to be determined. Beyond the international political and economic exigencies that have driven recent developments in water policy, there is still considerable scope for contestation and debate. If filtration of the city's entire water system does eventually occur at some point in the twenty-first century, historians of the future may well comment on the remarkable persistence of this particular fragment of engineered nature.

The extensive dam- and reservoir-building program undertaken by New York City caused wide-ranging disruption to the communities of the Croton and Catskill watersheds, yet the extending ecological frontier of the city enabled a new kind of mediation between nature and society that was of inestimable benefit to millions of people. Municipal-led policy interventions under the auspices of technological modernism have often had deleterious environmental consequences, as the highway-spliced inner-city neighborhoods of postwar New York attest (see chapter 3). Yet to dispense with the role of government altogether as part of an ecological critique of the institutional basis to Western modernity risks the effective abandonment of any practical means for implementing environmental regulation. This political and ecological dilemma is heightened by the global dimensions to environmental change, which are driven by the relative absence of any form of effective international economic regulation in the face of unprecedented capital mobility. The fact that global climate change may affect the hydrological conditions for New York's water supply in the future illustrates this profound uncertainty.[141]

Some recent critical interventions under the auspices of the postmodernity debate have tended to denigrate or least display profound ambivalence toward the regulatory role of government in modern societies. The demise of the nation-state is both predicted and welcomed as part of a new fluidity in cultural and economic life. "On the ethical front," writes the anthropologist Arjun Appadurai, "I am increasingly inclined to see most modern governmental apparatuses as inclined to self-perpetuation, bloat, violence, and corruption."[142] This kind of anti-statist or even conspiratorial sentiment is a recurring motif in environmental histories that are critical of urban demands on rural water resources. In an American context, for example, the water wars of the Midwest and southern California have proved a fertile ground for what we might term antimodern interpretations of large-scale state-directed water projects.[143] What is often missing from these accounts, however, is a fuller picture of the impact of the modernization of water infrastructures on the everyday lives of urban citizens. The provision of water remains a collective service, even if the public-private distinction has become sharper in recent years and even if the very word "public" misleadingly elides dominant economic, political, and cultural developments to the exclusion of more marginal voices. Water is a collectivity in a metabolic sense because urban life depends on its supply, but decisions over water policy have never been open to much in the way of public deliberation or debate. The most promising solution to environmental degradation may lie in the development of a more sophisticated public sphere through which new forms of democratic decision making can emerge in preference to any lurch toward the ecological Hobbesianism of greater control, which may prove in any case to be fiscally and ideologically untenable.

The role of strong advocacy groups and an informed and active citizenry emerge as crucial in any effort to protect the environmental advances of the past. Yet the post-New Deal environmentalist agenda harbors innate weaknesses: its individualist and consumer rights-based orientation serves to deflect attention from more widely conceived regulatory goals in the public interest which extend to the sphere of production as well as consumption (a dilemma we return to in chapter 5). Similarly, the degree of indifference on the part of city-based water quality advocacy groups to the economic viability of low-income upstate rural communities is testament to wider class-based tensions in the American environmental movement, which serve to strengthen the hand of capital in the dismantling of the public sphere. This political dilemma is heightened by the socially regressive consequences of market-led ecological modernization, epitomized by spiraling water charges, that threaten to fragment the political strength of any cross-class environmental alliances in the city.

This chapter has explored the evolving interaction between water and the dynamics of capitalist urbanization. We have seen how the creation of urban infrastructure has been essential to the economic viability of New York City and at the same time has fostered possibilities for new kinds of mediations between nature and society. The period from the 1840s until the 1970s marked a *longue durée* in the history of the city's water marked by a high degree of political and organizational continuity in spite of rapid urban growth and far-reaching technological change. The partial unraveling of existing relationships between water and urban form since the 1970s reveals the fragile and contradictory dimensions to the built environment within the ongoing process of capitalist urbanization. A precarious balance among disparate political, economic, and cultural understandings of urban water supply systems has begun to disintegrate. The creation of metropolitan nature necessitated immense technical and organizational ingenuity in order to link the hydrological cycle of upstate New York to a multiplicity of private spaces within the city. The experience of the last twenty years has revealed how the prospect of a disintegrating and contaminated water system has exposed deep anxieties over the state of the public realm.

2

Symbolic Order and the Urban Pastoral

The supply of the city with pure water was the noblest labor; the gift of its great lungs, or breathing-place, the next.

—*Henry Cleaveland*[1]

To conceive of New York without the park is to imagine the intolerable.

—*John Reps*[2]

An 843-acre strip of nature cuts through the heart of Manhattan Island. This is Central Park, widely considered to be the most important public space created in nineteenth-century America. The quiet northern edge of the park borders Harlem just a few blocks away from the abandoned and burnt-out lots which serve as a poignant reminder of the severity of postwar urban decline. At the southern end of the park we enter a different world: tourists and yellow taxis swarm around the upmarket shops and restaurants spreading south from Columbus Circle and Grand Army Plaza into the heart of the midtown business district. This streak of green connects two different worlds into a symbolic whole where the innate heterogeneity of urban life is forged into a unified realm. The

2.1 Aerial view of Central Park looking north, June 1938.
Source: McLaughlin Air Service. Courtesy of the New York City Parks Photo Archive.

city's exclusions and delights are thrown together in a tangled mass of human interaction. Here the tensions and contradictions of capitalist urbanization are softened under the shade of oaks, maples, and the remnants of the luxuriant vegetation that once covered Manhattan Island. The park now stands as a testament to the enduring place of nature in urban design. Lakes, meadows, and woods are traversed by a network of paths and bridges, combining formal elements with a variety of seminatural features. Looking beyond the park to the Manhattan skyline, one is powerfully aware that the whole landscape is one of human artifice, yet the presence of nature affords the semblance of continuity with the "first nature" of Manhattan Island.

This chapter explores the combination of political, economic, and cultural developments that contributed toward the creation of Central Park. We trace how the park has emerged as a focal point for a myriad of debates and controversies in the field of urban planning. The "greensward plan" devised in 1857 by Frederick Law Olmsted (1822–1903) in collaboration with the English architect Calvert Vaux (1824–1895) has entered the pantheon of urban design as one of the most innovative contributions to urban planning ever conceived.[3] Yet despite the lasting significance of Olmsted's contribution, the precise meaning of his design legacy is laced with ambiguity: his ideas have been appropriated at different times toward the cause of Fourierist utopian socialism, various permutations of American nationalism, both the endorsement and the rejection of Jeffersonian agrarianism, and more recently as part of the emerging environmentalist critique of "industrial society."[4] Since the 1950s, Olmsted's work and ideas have enjoyed something of a renaissance, reflecting a decisive set of changes in American planning in response to the excesses of technological modernism and the suppression of nature within urban design.[5] A typical recent intervention in the Olmsted debate is by the landscape architect Anne Whiston Spirn, who writes: "Much of Olmsted's work, written and built, is remarkably fresh a century after his retirement, but its potential has not been fully explored and realized. . . . Olmsted's legacy needs reclaiming."[6] But what is implied here by such a return to Olmsted? In what ways can nineteenth-century conceptions of urban design, however innovative at the time, usefully contribute toward the planning dilemmas of the twenty-first century?

This chapter considers how an uncritical reading of Olmsted has served to perpetuate narrowly formalist conceptions of urban form in combination with a series of crudely behavioralist interpretations of the interaction between nature and urban society. But it is not only the design elements of Olmsted's legacy that present difficulties for interpretation: his associated political ideas and his conception of the relationship between public space and the development of a democratic public sphere also warrant closer examination. In order to make better sense of the meaning and significance of Central Park, we need to differentiate between the park as a material embodiment of human imagination and its role as a symbolic representation of abstract normative ideals connected to the ideological legitimation of American society. Recent debates on the meaning of public space from within the tradition of neo-Marxist urban theory have tended to invoke a Lefebvrian distinction between spaces created through their use and the use of space for the imposition of social order.[7] While this distinction serves as a valuable heuristic device for developing a critical frame of analysis, this chapter will seek to avoid an unnecessarily dualistic conception of public space by insisting on the centrality of the ideological dimensions to the simultaneous evolution of public space as both symbol and material embodiment of the contradictory character of capitalist urbanization and the social production of nature.

Much scholarly attention has been devoted to understanding the role of Olmsted both as the park's principal designer and also as a pioneer within the emerging fields of urban planning and landscape design in nineteenth-century America.[8] In many cases, however, critical writings on Central Park have systematically exaggerated Olmsted's role to provide a highly individualist and often hagiographical interpretation of his legacy. Irving Fisher, for example, writes of Olmsted's "creative genius" and locates his landscape ideals at the leading edge of a Hegelian unfolding of urban history:

> Combining the elements of beauty, function, and morality, intrinsic to the aesthetic theories of German romanticism and American

> Transcendentalism, the creation of parks and city planning emerged as part of the reform movement of the latter nineteenth century. And the city planner became the artistic genius who, by the infusion of his thought into nature, recreates nature at a higher level of organic unity.[9]

For the historian James Fitch, Olmsted was one of a number of "visionary engineers" who were able to realize their projects at a unique juncture of social transformation during the middle decades of the nineteenth century. Olmsted, along with other leading designers and engineers such as Joseph Paxton, John Augustus Roebling, and Gustave Eiffel, are singled out as exceptional individuals who undertook tasks that "extended far beyond what would be today's ideas of professional responsibility."[10] In such commentaries we find nineteenth-century romanticism fused with the technical achievements of civil engineering and urban design to provide a remarkably individualized and aestheticized interpretation of urban form. Even Olmsted's codesigner Calvert Vaux caustically remarked in 1865 that the park would become merely "an ornament among many ornaments in the watch chain of Frederick Law Olmsted."[11] This chapter aims to unravel mythical conceptions of Olmsted's design legacy in order to emphasize the material circumstances behind the park's creation.

2.1 Cultural Anxiety, Land Speculation, and Public Space

The first serious attempt to conceive the future layout of Manhattan Island was set out in the well-known gridiron plan of 1811, in which undeveloped spaces beyond the edge of the city were portrayed as a checkerboard of individual land parcels awaiting investment. In reality, this plan was no more than a cartographic abstraction, since only the southern tip of Manhattan Island had yet been developed.[12] This original plan for the city envisioned only a slight expansion in open space, and even this scant provision was largely ignored in the following decades, with only fragments remaining as Union Square, Tompkins Square, and Madison Square Park. Furthermore, many of the city's smaller parks and squares remained private spaces accessible only to nearby propertyholders. Before the creation of Central Park, the use of "public" space within the city was sharply divided: on the one hand, there was a patchwork of private squares and gardens jealously defended against the threat of public use; on the other hand, there were pleasure grounds or open spaces at the margins of the city, frequented by the city's working classes, in which music, games, and traditional festivals could be enjoyed.[13]

By the middle of the nineteenth century the urban landscape of Manhattan was one of increasing congestion, pollution, and social unrest, with serious riots in 1834, 1837, and 1849.[14] Deteriorating urban conditions set in train a complex series of debates among the city's political and cultural elites as to how the rapidly growing city could be improved. From the late 1840s onward we find increasing demands for a new public space in the city, articulated through newspaper editorials, political speeches, and articles in professional journals. Prominent early advocates for a large park in Manhattan include Walt Whitman, the editor of the *Brooklyn Eagle,* and William Cullen Bryant, the editor of the *New York Evening Post.* In an editorial entitled "a new park" in July 1844, for example, Bryant demanded the creation of "the public garden of a great city."[15] In the following year Bryant wrote from England to lament: "The population of your city, increasing with such prodigious rapidity; your sultry summers, and the corrupt atmosphere generated in hot and crowded streets, make it a cause of regret that in laying out New York, no preparation was made, while it was yet practicable, for a range of parks and public gardens."[16]

An early supporter of Bryant was the influential writer and landscape architect Andrew Jackson Downing (1815–1852). Writing in *The Horticulturalist,* Downing emphasized the role of nature as a civilizing influence on society and a path to knowledge and refinement.[17] He also saw parks as a means to foster the greater democratization of society through the establishment of "a larger and more fraternal spirit in our social life,"[18] a theme that was to prove pivotal to the success of Central Park as a symbol of both political and aesthetic advancement. For Downing, writing in 1851, the new park was to be emblematic of the known world, of all accumulated knowledge and learning, and a mark of the highest human achievement.[19] Hidden behind this kind of civic rhetoric was a counterdiscourse of social elitism, forcible education of the masses, and a kind of cultural and technological rivalry with the cities of Europe. Indeed, as we shall see, the debate over public space was not a public one in the sense that there was any real engagement between the city's social and cultural elites and the aspirations and concerns of the city's burgeoning working-class population and newly arrived immigrant communities.

Early lobbyists for the creation of a new park were mostly wealthy merchants and landowners who had admired the public spaces of London, Paris, and other European cities. Typical of these was the merchant Robert Minturn and his wife Anna Mary Wendell, who brought together a committee to lobby for a new park in the early 1850s. The absence of adequate public space in New York was widely remarked upon as a mark of nineteenth-century America's cultural backwardness in comparison with Europe (a sentiment strengthened by the observations of overseas visitors such as Alexis de Tocqueville).[20] The European design legacy had also been fostered in North America through a number of naturalistic designs for urban cemeteries, beginning with Mount Auburn Cemetery near Boston in 1831, followed by Laurel Hill Cemetery in Philadelphia in 1836 and in particular by Brooklyn's Greenwood Cemetery in 1838. These cemeteries served to popularize landscaped gardens among urban elites, who developed a taste for the English picturesque tradition in landscape design as well as for the extensive deployment of statues and water features redolent of European gardens.[21] Cemeteries as pockets of tranquil nature within rapidly expanding North American cities were to play a significant role in influencing elite opinion at a crucial stage in the emergence of early urban planning ideals.

In the wake of the European romantic tradition, contact with nature was widely conceived as a means to foster a civil society capable of emulating the ostensibly stable and ordered societies of the Old World (a rather ironic sentiment given the political turmoil of nineteenth-century Europe). Arguments in favor of nature in cities were also imbued with a belief in the curative and circulatory powers of green spaces. Parks were widely portrayed as "urban lungs" capable of countering the effects of disease-carrying miasmas by facilitating greater ventilation and movement of air. In this sense the parks movement found a wider resonance in the "hygienist" discourses of nineteenth-century public health reform. The *Irish American,* for example, insisted that the park be intended not for the rich "but as the 'lungs of the city' for the working classes."[22] The arguments for public space involved an interplay between premicrobiological theories of disease transmission and organic metaphors for the healthy functioning of cities. Olmsted himself was to claim in his memoirs that "air is disinfected by sunlight and foliage," adding a scientific veneer to the moral promotion of nature in cities.[23]

But the decisive argument in favor of parks, and the one consistently played down within the aesthetic concerns of the city's cultural elite, was the growing significance of real estate speculation in urban design. Owners of land and property in different locations across the city began to lobby to have the new park in their own vicinity. Earlier suggestions such as the planned extension to Jones Park on the Upper West Side provoked angry disagreements over how the proposed park should be financed. In order for a very large public space to be created, the potential benefits would have to be shown to accrue to the city as a whole rather than simply to local property owners, so that the costs might also be spread as widely as possible. In the event, a compromise was eventually reached over the financing of a new park by combining betterment assessments with new forms of general taxation.[24] The crucial significance of real estate speculation is borne out by the rapid impact of the park on the value of nearby land and property. Undeveloped lots at the corner of Fifth Avenue and 86th Street that were valued at little more than $500 in 1847 were worth in excess of $20,000 by 1868.[25] Descriptions of the park shortly after its completion reveal the extent of the economic advantages that had accrued to the city's property owners. In 1863, for example, Frederic B. Perkins reflected that "the Park now represents an expenditure of more than seven million dollars. . . . But against this debit of seven million dollars may be set a sum of *twenty-two and half millions* of dollars, being the increased value, since 1856, of the real estate in the three wards surrounding the Park, and of which a considerable part is due to the influence of the Park."[26] By 1868 New York had entered an inflationary speculative boom in vacant lots concentrated around the new park. With further improvements in building design, rapid transit, and other aspects of urban infrastructure, the land market in Manhattan had been irrevocably altered. We should note that the speculative craze was not restricted to New York but extended to Boston, Chicago, San Francisco, and other cities swept along by the land inflation of the post-Civil War years.[27] What we find in Manhattan is an intense interaction between local and national factors that contributed toward the escalation in land values.

The growing sophistication of real estate speculation in Manhattan could not have occurred without a series of legislative and administrative innovations in urban government. In May 1851 Mayor Ambrose C. Kingsland made the unprecedented demand that the city use its local tax base for the creation of a large public park that would serve as "a lasting monument to the wisdom, sagacity and forethought of its founders."[28] In 1853, the New York State Legislature authorized the City to use the power of eminent domain to acquire some seven hundred acres in the middle of Manhattan (a legal device that had already been employed for the construction of the city's water system); and in 1857, the State Legislature created the Central Park Commission, which was to control the park's construction and design. The Central Park Commission was the city's first planning agency and heralded an effective end to piecemeal, uncontrolled urban growth on Manhattan Island. Taken together, these various initiatives involved direct large-scale public intervention in the private land market on behalf of a new set of institutional arrangements to create "a space permanently removed from the private real estate market."[29] This is not to suggest, however, that these institutional arrangements worked against the interests of private capital. On the contrary: the park's creation altered the relationship between municipal government and private capital under the guise of a newly defined "public interest" within which the prospects for real estate speculation were greatly enhanced.

Opposition to the park was led principally by wealthy property owners who argued that the cost to taxpayers was too high, and that in any case the city lay within easy reach of surrounding countryside and open water. Other middle-class fears centered on crime and the opportunities for the congregation of undesirable social elements such as Irish and German "liquor dealers."[30] Even the progressives were uncertain: the New York Association for Improving the

Condition of the Poor feared that the park would force yet more overcrowding in the slum tenements of congested parts of the city.[31] The creation of a park commission appointed by New York State also provoked fierce political rivalries between Yankee Republican interests at state level who strongly backed the park and city-based Democratic supporters who were suspicious and resentful of what they perceived as outside interference in the management of city affairs. In May 1857, for example, some twenty thousand New Yorkers gathered in City Hall Park to denounce the Republican-dominated State Legislature's interference in local affairs and called for the proposed park to be abandoned.[32] The articulation of a putative general interest in the face of prevailing laissez-faire ideology was no easy matter: thousands of owners of property in or near the south side of the proposed park site demanded that its dimensions be reduced. Given the rapid physical expansion of the city and escalating land values spreading toward the undeveloped north of Manhattan Island, along with the projected cost of the project (at some three times the city's annual budget), the slightest delay might very easily have prevented the park from being built.[33]

A further uncertainty stemmed from cyclical fluctuations in the regional economy. In January 1854 the New York economy entered a sharp downturn that would lead into a depression and the financial panic of 1857. In March 1855 the deteriorating economic conditions led the city's Board of Aldermen to adopt the proposals of Mayor Jacob Westervelt to reduce the size of the park on the grounds of fiscal prudence. However, the election of Mayor Fernando Wood in 1855 reversed this; Wood argued that the creation of Central Park would be an "intelligent, philanthropic and patriotic public enterprise" and successfully cultivated support for the project among both wealthy merchants and also poor immigrant neighborhoods who would benefit from its use. In addition to shoring up his future political ambitions, Wood also stood to benefit personally from the project as a major West Side landowner: lots fronting the park that he had bought for a few hundred dollars were each worth ten thousand dollars by 1860.[34]

Three dominant themes emerge from these debates surrounding the creation of public space in mid-nineteenth-century New York. First, there is the emergence of a Yankee predilection for the English picturesque landscape transposed to an industrial urban setting, in the context of a pervasive cultural anxiety on the part of North American elites who consistently compared American cities with those of Europe. Second, at a political level the theme of the "public interest" was skillfully manipulated in order to impose a particular conception of urban order amid rapid and seemingly chaotic patterns of urban change. And third, the growing sophistication of real estate speculation becomes linked with newly emerging conceptions of the aesthetics of nature and urban design. This last theme is of particular interest in that it facilitated an uncanny degree of congruence between a distinctively American nature aesthetic derived from the legacy of romantic idealism and the emergence of a sophisticated metropolitan ideology of nature within which the commodification of nature as a social product became an integral dimension to the dynamics of capitalist urbanization.

2.2 Creating the Garden of a Great City

In 1857 the newly reelected Central Park Commission held a national contest for the design of the park, and the "greensward plan" devised by Frederick Law Olmsted and Calvert Vaux was chosen from some 33 competing entries.[35] The greensward plan combined three main elements: a pastoral landscape with open rolling meadows, exemplified by the so-called Sheep Meadow; a more naturalistic "picturesque" design, most obviously represented in the semiwild landscape to be found in the Ramble; and a variety of formal elements contained in the fountains, lakes, and boulevards of the Mall, the Promenade, and Bethesda Terrace. A further unique feature was that all four roads traversing the park were to be constructed at a lower level than the park's surface in order to emphasize the sense of an uninterrupted green space in the center of the city. In terms of design precedents, the greensward plan drew primarily on a naturalistic design and notably eschewed contemporary alternatives such as the crudely utilitarian geometries of "republican simplicity" epitomized by the gridiron street plan; the largely unplanned popular eclecticism to be found nearby in the pleasure gardens of Hoboken's Elysian Fields; and in particular the imperial and neoclassical tropes of artificial civic display best known to Olmsted and Vaux through the reconstruction of Second Empire Paris under Napoleon III.[36] Above all, the park was to be unequivocally separate from the rest of the city not just aesthetically but also in its culture of use, to create a new kind of public space for a more refined conception of American urban life (figure 2.2).

The construction of Central Park involved some twenty thousand workers for the removal of three million cubic yards of soil, the planting of over 270,000 trees and shrubs, and the building of a new reservoir for the city's recently completed water supply system.[37] Much of the land purchased for the park consisted of swampy and rocky plots undesirable for private development. Northern parts of the site were already a city landmark on account of their oak, chestnut, and elm forests: a fragmentary reminder of Manhattan Island's former beauty that was to be incorporated into the park's design.[38] Toward the southern end of the site, however, lived some of the most marginalized communities in the city (figure 2.3). The erasure of these communities may even have been a significant motivating factor behind the political momentum for the park's creation (and the fear that informal settlements would proliferate). Consider, for example, the description of the original site offered by Perkins in his popular guide to the new park:

> In various portions of its savage territory, tribes of squalid city barbarians had encamped, and, in dirty shanties or in the open air, drove the fetid business of bone-boiling—"dreadful trade;" nourished herds of measly swine upon the sickish feculence of distilleries, or murdered rapid successions of wretched "stump-tail" cows, who dissolved bodily into mere rottenness on the same nauseous food, as they stood in the stalls, poisoning the city infants with their infectious milk as they died. Cinder-shifters, rag-pickers, and swill-men constituted its more cleanly or aristocratic classes, unless now and then some thief or bolder criminal glorified its huts or holes with a more famous presence. It was a miserable realm of barrenness, stench, filth, poverty, lawlessness, and crime.[39]

2.2 Central Park, looking north from 59th Street, lithograph by C. Bachman (circa 1865). Note the empty lots depicted at the edges of the park.
Source: J. Clarence Davies Collection, Museum of the City of New York.

2.3 Squatters' shacks in the vicinity of the Central Park site (1862).
Source: Collection of the New-York Historical Society. Victor Prevost/George Eastman House, Rochester.

Surviving historical sources suggest that the construction of the park displaced some sixteen hundred residents of shantytowns in the designated area, including Irish pig farmers, German gardeners, and what may have been the city's most significant antebellum black settlement called Seneca Village, which included three churches and a school.[40] The destruction of Seneca Village illuminates the degree to which the "public interest" involved not only the erasure of existing communities but also the promotion of a particular conception of a unified urban society and an intense marginalization of those groups that fell outside this conception.

The disparate influences on the greensward plan illuminate the interconnections between the park's design and a broader ideological agenda emerging at a unique juncture in American history. Olmsted and Vaux explicitly rejected the suggestions of Richard Morris Hunt for ornamental gates and other imperial symbols reminiscent of Haussmann's Paris.[41] In this sense they sought to construct a distinctively republican aesthetic capable of meeting the aspirations of a far wider public than had hitherto been served by American landscape design. They also eschewed Andrew Jackson Downing's design for the park with its emphasis on "forcible education" through labeled trees and shrubs.[42] In rejecting Downing's approach they distanced themselves from the most paternalistic and narrowly didactic dimensions to nineteenth-century social reform. Yet Downing's promotion of English landscape gardeners such as Humphry Repton, John Nash, and Joseph Paxton was to prove influential, not least through Vaux's earlier professional association with Downing (and the fact that Downing had been one of the leading American exponents of romantic landscape design before his death in 1852).[43] We know from Olmsted's writings that the design legacy of Uvedale Price, William Gilpin, and Humphry Repton proved highly significant. Olmsted had also enthused over "picturesque" park designs he had observed in England during his visit to Europe in 1850, such as Stowe, Stourhead, and Blenheim.[44] The intellectual and aesthetic lineage between Repton and Olmsted is especially intriguing because Repton expanded existing conceptions of landscape improvement to encompass changes in social relations, to the chagrin of his contemporary antagonists such as Price and Richard Payne Knight. The modern civic vision of

Repton finds resonance in Olmsted's search for a more democratic republican landscape in the face of tensions between elite and popular tastes, uneasy relations among different bodies of knowledge and professional expertise, and conflicting conceptions of national identity and landscape iconography.[45]

The aesthetic theories of Price and Gilpin also provided an alternative model to the more formal approaches exemplified by Adolphe Alphand's "manicured" designs for the Buttes-Chaumont, constructed during the same period as part of the imperial facade of Second Empire Paris.[46] The artist and critic Robert Smithson has argued that this intellectual lineage to Price and Gilpin forms part of a radically different conception of nature to that derived from the German romantic tradition:

> As a result we are not hurled into the spiritualism of Thoreauian transcendentalism, or its present day offspring of "modernist formalism" rooted in Kant, Hegel, and Fichte. Price, Gilpin, and Olmsted are forerunners of a dialectical materialism applied to the physical landscape. Dialectics of this type are a way of seeing things in a manifold of relations, not as isolated objects. Nature for the dialectician is *indifferent* to any formal ideal.[47]

While Smithson is right to emphasize the ambiguity of any direct connection between Olmsted's aesthetic and the legacy of European romanticism, we should not overlook the blurring of aesthetic and political judgment that links Olmsted's conception of the civic realm to the Enlightenment preoccupation with the refinement of taste (and the concomitant distrust of popular culture).[48] We can also find a degree of continuity between Olmsted and the pantheistic romantic vision of Ralph Waldo Emerson, Henry Thoreau, Walt Whitman, and Herman Melville. The creation of Central Park can be perceived as part of the emergence of a distinctively American tradition of neoromantic nature aesthetics in the context of the irreversible transformation of American landscape and society.[49] For Leo Marx, it is Emerson and Whitman who provide the most vibrant celebration of "this industrialized version of the pastoral ideal," as a republican fusion of nature and culture in the new urban America.[50] Central Park provides a dramatic illustration of what Marx calls the "middle landscape," which is necessarily dialectical in relation to the Jeffersonian rural idyll and its evolving relation to nineteenth-century urbanization.[51]

What was distinctive about the emerging dialectical approach to landscape design in nineteenth-century America was the articulation of a new kind of mediation between nature and culture that self-consciously evoked a metropolitan nature aesthetic. The creation of Central Park marked a growing aesthetic distinction between landscapes for production and for consumption: beyond the park lay the increasingly rationalized agricultural landscapes and more distant plantations and trade networks that sustained the transformation of urban space; within the park, an imaginary natural order existed as a new form of cultural consumption emanating from emerging patterns of touristic engagement between the urban middle classes and the perceived wilderness of "first nature." Evidently, some visitors were perplexed at how much the park resembled their preconceived conception of a wild landscape rather than a meticulously designed urban park:

> Even the Park itself has been somewhat of a disappointment, according to the preconceived ideas of the visitors. There are those who look for great, sculptured gateways, and a scene of fountains and statues, and to whom the passage through the openings in the low wall will seem but a going out into the country. The remark of Horace Greeley on his first visit—"They have let it alone more than I thought they would"—comes to these in a different sense; and it is only by remembering the wilderness of rocks and shanties, stagnant pools, and bare, rubbish-strewed soil, on Fifth, Seventh, and Eighth Avenues, in these high latitudes, three years ago, that one can realize the wonders that have been worked in making this *seeming* Nature what it is.[52]

This phrase "*seeming* Nature" is crucial to any understanding of the cultural and ideological significance of the park as an imaginary representation of nature for a sophisticated urban audience. The semiwild features of Central Park are most strikingly represented in the Ramble (figure 2.4), which resembled the kind of scene that was being popularized by the growth of tourism to mountainous areas during the early nineteenth century. The creation of the Ramble reflects the emergence of a distinctively American approach to landscape aesthetics associated with Thomas Cole and the Hudson River School.[53] These artists played a key role in popularizing "wild nature" in the early decades of the nineteenth century, and their work would certainly have been familiar to many New Yorkers who craved something of the "cool country" in the midst of the summer heat.

In Cole's *The Oxbow* (1836), however, we can detect a tension between two contrasting types of landscape iconography: an imaginary wilderness of "first nature" and a very different kind of nature aesthetic derived from the classical trope of cultivated gardens and riparian civilization (figure 2.5). In Central Park a similar contrast divides the imagery of the Ramble from that of the Sheep Meadow, drawing on different kinds of cultural engagement with nature. The "urban pastoral" element of the Sheep Meadow can be interpreted as an allegorical form masking the demise of the rural Jeffersonian ideal.[54] At the very moment when ostensibly stable agricultural societies were being irrevocably transformed, an aesthetic veneer of rural imagery was being busily recreated in urban space. In this case, the landscape has become a kind of fetishized commodity, the high point of an "agrarian bourgeois art" described by Raymond Williams as "a rural landscape emptied of rural labour and of labourers; a sylvan and watery prospect, with a hundred analogies in neopastoral painting and poetry, from which the facts of production had been banished."[55] Just as the sunken roads are artfully hidden from view, so the real relationship between the park and the city is difficult to discern. The role of the human hand is rendered uncertain, leaving the park with no apparent origins. The park workers remain largely invisible, holding the landscape in a suspended animation of ecological succession for the aesthetic adornment of the city.

2.4 Central Park's Ramble in Fred. B. Perkins, *The Central Park* (1863). View north toward cave showing plantings of pine and larch trees.
Source: Avery Architectural and Fine Arts Library, Columbia University.

2.5 Thomas Cole, *View from Mount Holyoke, Northampton, Massachusetts, after a Thunderstorm (The Oxbow)* (1836).
Source: Courtesy of the Metropolitan Museum of Art, New York.

The design of Central Park tells us much about changing perceptions of nature in nineteenth-century American thought. Its imaginative combination of so many seemingly irreconcilable elements makes it a kind of medley of different aesthetic responses to capitalist urbanization. But our consideration of park design leads us to wider questions concerning the relationship between public space, civil society, and the emergence of new forms of urbanism. How, for example, did the park design relate to a broader concept of urban society and the possibilities for cultural and political advancement? And how did the urban vision engendered by the creation of Central Park actually connect with the changing political, social, and economic complexion of the rapidly growing city?

2.3 Olmsted's Urban Vision: A Fragile Synthesis

To understand the paradoxical dimension to the design for Central Park, we need to examine Olmsted's conception of the "ideal city" as a distinctive alternative to both the violent anarchy of the Western frontier and the repressive backwardness of the rural South.[56] For Olmsted, writing in 1858, the park was "a democratic development of the highest significance and on the success of which, in my opinion, much of the progress of art and esthetic culture in this country is dependent."[57] This is a revealing and important claim because it shows how Olmsted's landscape vision combined political and aesthetic concerns in an urban context. But how did politics and aesthetics interact in antebellum New York? And how might the creation of a new public space actually contribute toward the advancement of democratic ideals and more refined conceptions of urban life?

Olmsted's concerns with urban design must be placed within the context of a series of decisive changes in nineteenth-century American society. The historian Albert Fein suggests that the decline of slavery and the concomitant emergence of new democratic and ethical ideals suffused Olmsted's sense of political responsibility to build a better kind of society. In his vivid description of the antebellum South—*The Cotton Kingdom*—Olmsted argued against slavery and the plantation system on the grounds of justice and natural law.[58] Yet Olmsted's conception of social change was cautious and incremental, combining ethical and economic arguments in a kind of pragmatic synthesis.[59] Olmsted was especially concerned with the effects of slavery on the economic efficiency of American agriculture and, far from condemning southern elites, wrote of his admiration for "true and brave Southern gentlemen."[60] We can argue that Olmsted was no radical or abolitionist, but a paternalist whose writings on the South illuminate the complex political and economic tensions that would explode in the Civil War.[61] The particular significance of his antipathy toward the rural South lay in his identification of cultural advancement with the growth of cities freed from the strictures of an agrarian economy. Olmsted's conception of modern society was founded on a harmonious interplay between nature and culture within which cities, with their civic institutions, cultural vibrancy, and ideas, formed the "natural fruits of democracy."[62] Central to his vision was a belief that if only the physical hardships of nineteenth-century urbanism could be overcome, the real potential of urban life might be realized. The city was not an aberrant social form to be feared but a dynamic focus for the creative energies of modern society.

The combination of republican ideals with urban design can be traced to Olmsted's visit to England's Birkenhead Park in 1850. Enthused by what he described as "this people's garden," Olmsted was determined to produce a distinctively republican landscape in an American context.[63] Joseph Paxton's Birkenhead Park, which opened in 1847, was the first state-directed park to be constructed in England, as distinct from the royal parks of the past. This political distinction is borne out by contemporary responses to Birkenhead Park as "a great democratic pleasureground; a proof of the ease and the natural method by which a democracy can create, for its own enjoyment, gardens as elaborate, costly, and magnificent, as those of monarchs."[64] Given Olmsted's concern with republican inclusiveness in urban design, it is somewhat ironic that it was the growing tension between aesthetic elitism and popular culture that would ultimately undermine his direct involvement with Central Park and lead to the marginalization of the political and cultural values he espoused.

The articulation of a nature aesthetic for a wide public audience marks a significant break from the European intellectual heritage that found its apotheosis in the work of Andrew Jackson Downing. American cultural elites of the antebellum era were disdainful toward what they characterized as southern European "indifference" to the refined aesthetics of nature embodied in the northern European romantic tradition. By associating nature appreciation with northern European culture, Downing and others had adopted an explicitly racialized aesthetic in which the appreciation of the beauty of nature was considered to be an Anglo-Saxon domain. For Susannah Zetzel, however, this contrasts with Olmsted's recognition of the social heterogeneity of urban America and the futility of pursuing an elitist aesthetic: "Where Downing thought that only the Anglo-Saxon races could be ennobled by contact with nature, and that for others more explicit direction would be necessary, Olmsted's vision was universal."[65]

Olmsted shared with many other American intellectuals a concern with the need to create a unifying national culture in the midst of sweeping social and economic change. Mass immigration was creating a complex urban society in which any semblance of Anglo-Saxon cultural homogeneity was fast disappearing: by 1855 some 52 percent of New York's population was foreign born. Early responses to the park repeatedly emphasized the rhetorical theme of the whole spectrum of urban society brought together, even if the reality was very different. In 1861, for example, Charles Eliot Norton claimed that Olmsted stood "first in the production of great works which answer the need and give expression to the life of our immense and miscellaneous democracy."[66] But what kind of public space was envisaged by the construction of Central Park? Can we really conceive of Olmsted's vision as universal?

The park was certainly never intended as a forum for political debate and the promotion of discursive interaction between strangers. It was rather an enlargement of the private sphere through the extension of nineteenth-century conceptions of bourgeois domesticity into a public arena. If we examine the way the park was managed, we find that the use of this newly created public space was initially highly restrictive, with rules to discourage picnics and other group activities.[67] Despite these restrictions, public use of the park grew rapidly: when the park finally opened in the winter of 1859, thousands skated on lakes built over the former swamps (figure 2.6). By 1863 the park was receiving more than four million visitors a year, rising to over seven million visitors in 1865 and

2.6 *Central-Park, Winter. The Skating Pond,* lithograph by Currier and Ives (1862). Source: Print Collection. Miriam and Ira D. Wallach Division of Art, Prints and Photographs. Courtesy of the New York Public Library. Astor, Lenox and Tilden Foundations.

nearly 11 million by 1871.[68] In its early years, however, the park was effectively an elite playground, with access largely dependent on the use of carriages (which only a small fraction of the population could afford). Much was made of the social display of early park users and especially the fashions sported by women park visitors at a time of newly emerging urban spaces for leisure, consumption, and ostentatious displays of wealth.[69] In subsequent decades the development of public transport improved accessibility for poorer and more distant parts of the city (the impressive IRT subway only began running in 1904) and the restrictions on park use were gradually relaxed, leading to peak usage in the early decades of the twentieth century before rising levels of car ownership led to new patterns of urban leisure.

The relationship between park design and the recreational needs of a wider public than that envisaged in Olmsted's original plan was to prove a pivotal element in political disagreements surrounding the control of Central Park. Olmsted's refusal to include sports facilities in the park design left him vulnerable to allegations of disdain for the popular culture of industrial America. In 1859, for example, he insisted that "no sport can be permitted which would be inconsistent with the general method of amusement."[70] The populist city-based political machines were quick to emphasize any elitist connotations in Olmsted's design. The "genteel reformers," with whom Olmsted had some association, were opposed to political and economic change and sought to maintain the dominance of an educated elite.[71] The cultural elitism of Olmsted and his followers left them politically vulnerable to newly emerging populist machine politics, particularly the Irish Democratic power vested in Tammany Hall from the late 1860s onward, which naturally viewed English cultural traditions with suspicion if not disdain. Tammany press outlets such as the *New York Evening Express* ridiculed Olmsted's backers as "the Miss Nancies of Central Park art" who "babble in the papers and in Society Circles, about aesthetics and architecture, vistas and landscapes, the quiver of a leaf and the proper blendings of light and shade."[72] The Yankee elites that had fostered the park's development in the 1850s found their conception of the "public interest" rapidly marginalized, leading to Olmsted's dismissal in 1878 from any managerial role in the park. From the 1870s onward a new phase of capitalist urbanization fostered a different set of political and economic relations.[73] The emerging political polarization between increasingly militant labor organizations and the power of capital tended to sideline the aesthetic and didactic concerns of Olmsted and his followers. Over time, the fragility of his conception of democratic society became exposed to critical scrutiny, yet his influence on landscape design continued to grow: a paradox that permeates contemporary responses to his work and ideas and invites a closer consideration of the enduring power of nature within urban design.

2.4 Olmsted Rediscovered: An Emerging Preservationist Ethic

For Olmsted, urban planning was an art in which aesthetic concerns took precedence over any more radical criticisms of the workings of the urban land market or of society more generally. The greensward plan for Central Park was an aesthetic vision imposed on society as an explicit alternative to what he perceived as the "crude and materialistic impulses of popular culture."[74] Though he conceived of public space as the physical embodiment of a democratic society, the design and management of these spaces was to be left to a technical elite, suggesting a profound ambivalence toward more concrete forms of democratic participation. Olmsted's approach epitomized an emerging characteristic of urban planning, as "cosmopolitan élites, deprived of grass-roots political power, learned to assert their authority in public life through specific expertise in the higher echelons of urban governance."[75] As the historian Thomas Bender notes of Olmsted and his allies, they "were concerned to establish their opinion in public; they were not interested in a public or political sphere that served as an arena for competing ideas and interests."[76] The overwhelming priority of nineteenth-century philanthropists was one of order: a social and spatial order within which the interrelated problems of "pauperism, congestion, environmental chaos, and aesthetic disarray" could be handled by a combination of professional expertise and advances in scientific knowledge.[77] The promotion of urban parks can be conceived as simply one element of "the larger social-improvement crusade" that developed in nineteenth-century America.[78] In this sense, the nineteenth-century urban park forms an integral element in the emergence of a distinctive political vision rooted in a redefinition of relations between nature and culture.

For the artist and critic Robert Smithson, Olmsted threw "a whole new light on the nature of American art" with his distinctively dialectical view of nature and society.[79] But can we really conceive of Olmsted's vision as "dialectical" in a Marxian or Lefebvrian sense? Or is the design for Central Park best interpreted as an ambiguous and transitional moment between urban beautification and the emerging dynamics of technocratic planning that would develop rapidly from the late nineteenth century onward? Olmsted's conception of public space developed out of a dual emphasis on the maintenance of social order and the enrichment of civil society. In this sense, one could argue that Central Park embodied the Lefebvrian tension between spaces appropriated through their use and spaces devised in order to impose order. It is difficult, however, to apply the Lefebvrian conceptual schema directly to Central Park without qualification. The park has passed through a series of contrasting phases in its use and meaning. In the era preceding the park's construction, much of the chosen site had already become a public space through extensive cultivation, grazing, informal settlement, and a variety of marginal economic activities. When the park opened it became an ongoing focus for markedly different conceptions of civic culture, recreation, and social interaction, which effectively superseded the original rationale for the park's creation.

More recently, a period of relative openness in the park's political and public role has given way to a revanchist reiteration of the original vision under the guise of an increasingly privatized approach to park management and control. From the late 1960s onward, for example, we find a rediscovery of Olmsted marked by a shifting emphasis in park management initiated under Mayor John V. Lindsay toward the preservation and enhancement of Olmsted's original designs. Since the 1980s, an Olmstedian conception of order and harmony has been repeatedly invoked in order to provide an "aesthetic" solution to the contradictions of capitalist urbanization in the post-Fordist era. This renewed commitment to the aesthetics of park management has coincided with the growing manifestation of poverty, destitution, and homelessness in New York's most important public space.

Whereas the nineteenth-century political rationale for state intervention in urban design rested principally on the twin concerns of public health and social order (however ill defined), in the late twentieth century the focus moved inexorably toward crime and combating the consequences of social exclusion. What is striking is the degree of ideological continuity between the original pretext for the park's creation and the contemporary emphasis on the more regularized and stylized dimensions to Olmsted's urban vision. Successive park management plans have focused on the restoration and recreation of original features in the Olmsted design as part of a new commitment to the preservation of urban architectural heritage.[80]

The Central Park Conservancy, set up as a private charity in 1980, now provides half of the park's operating budget, provides half the funds for capital projects, and employs around eighty percent of the park's staff.[81] This organization has raised more private donations for a public park than any other in American history, transforming Central Park over twenty years from a "graffiti-smeared ruin that was an international embarrassment" into a "meticulously restored greensward." To do this, the Central Park Conservancy has tapped into complex social networks of wealthy patronage and charitable giving in the affluent wards bordering the park and has enlisted the expertise of leading financiers such as the billionaire leveraged-buyout tycoon Henry R. Kravis.[82] The restoration of the greensward vision has reinforced the vast disparities in wealth and land values associated with the completion of the park in the nineteenth century. Since the park budget cuts of the 1970s, there has been growing inequality in access to and quality of public space across the city. By the early 1990s New York ranked nineteenth among major American cities in terms of per capita public expenditure on its park system (far below Los Angeles or Chicago, for example) and was left with just half the park staff it had in 1960.[83] Recent fund-raising successes for Central Park may be contrasted with neglected municipal parks elsewhere in the city. Some illustration of the uneven capacity to raise private money for public space is provided by two fund-raising dinners held in 1999: a dinner held for Marcus Garvey Park in Harlem raised $7,000 whereas a Central Park event raised $805,000.[84] Even the possibility of spreading the wealth of Central Park to other Olmsted-designed parks such as the shabby Morningside Park in Harlem has been resisted by donors: it seems that the commitment to restoring Olmsted's pastoral vision is highly localized, not a citywide objective. The park has become a cultural institution in its own right comparable with the Museum of Modern Art, the Metropolitan Museum, and other symbols of social and political prestige in the city.

The 1990s saw a sharp polarization of perspectives on public space, with community gardens and other alternative "green pockets" under intense pressure from developers despite well-organized opposition from grassroots organizations such as the Green Guerrillas and the Neighborhood Open Space Coalition.[85] The scale of gentrification pressures on urban space has intensified since the 1980s, producing new inequalities driven by international as well as national flows of capital and investment. The temporary lull in the wake of the 1987 stock market crash and the recession of the early 1990s has been superseded by a new wave of development pressures far greater than in the past.[86] Central Park is now the tranquil core of an increasingly globalized dynamic of land commodification in Manhattan. It is striking that the pressures that have contributed toward Central Park's cultural and financial renaissance have simultaneously undermined the prospects for the city's hundreds of community gardens concentrated in poorer neighborhoods (figure 2.7). This example should discourage any simplistic connection between public space and the promotion of urban nature as a dimension to ecologically framed conceptions of urban design.

As we have seen, Olmsted's vision was never a public one from its inception but that of an urban elite who successfully imposed it on the wider society. The fact that Central Park has subsequently acquired a historical and cultural legacy of public action and collective memory is incidental to its original rationale within the commodification of both land and nature in the middle decades of the nineteenth century. Nonetheless, the park has been used to illustrate an abstract, normative ideal of an inclusive public sphere that is held to have existed in the past.[87] In *City of Quartz,* for example, Mike Davis invokes Olmsted's Central Park

2.7 The Magnolia Tree Earth Garden, Bedford-Stuyvesant, Brooklyn, 1996.
Source: Photograph by Matthew Gandy.

as America's answer to the class polarization of nineteenth-century Europe and a symbol of a more progressive social order.[88] Davis suggests that the Olmsted legacy provides an alternative to the militarization of urban space in contemporary America. Yet the fact that Olmsted's original ideas have been drawn on in recent years to shore up a preservationist and in many respects exclusionary conception of public space reveals the precariousness of any argument that relies on nineteenth-century conceptions of urban form as a means to challenge the contemporary gentrification of public space. Recent debates on the future of public space have rightly questioned a nostalgic attachment to the illusory public spaces of the past.

In the case of Central Park, its zenith in terms of social inclusiveness was probably reached in the Lindsay era of the 1960s and early 1970s with new festivals, happenings, and large-scale political gatherings (figure 2.8).[89] Yet the new openness to different uses of public space by Mayor Lindsay and his Parks Commissioner August Heckscher provoked fierce criticism from wealthy residents bordering the park and from conservative politicians who objected to "flag burning rights in Central Park."[90] This period also coincided with the emergence of a combination of fiscal and managerial problems in park maintenance that gradually assumed increasing significance in the politics of public space. The contemporary emphasis on "reclaiming" the park and making the space more attractive to middle-class New Yorkers is as much a reflection of socioeconomic shifts within the city as of any rediscovery of Olmstedian ideals in urban planning. The fact that Olmsted's aesthetic ideals find a strong resonance with an emerging preservationist ethic reveals the ease with which ostensibly progressive social ideals have been transmuted into new forms of cultural boosterism to serve the needs of powerful urban elites. Recent changes in park management and funding may have improved the park's appearance, but the "sovereign public" has seen an erosion of control over the city's most important public space.[91] To rely on Olmsted's legacy to chart a coherent critical response to the precarious status of contemporary public space not only risks a simplification of the relationship between public space, the public sphere, and development of civil society, but also

2.8 Anti-war "Love-Lee," Sheep Meadow, Central Park, circa 1968.
Source: Museum of the City of New York.

perpetuates an anachronistic nineteenth-century frame of reference for understanding the place of nature in urban design.

To return to the influential neo-Marxian interventions by Henri Lefebvre, which have proved a recurring element in recent debates over public space, we should note that Lefebvre has a weakly developed theorization of the social production of nature, which has been a crucial ideological and aesthetic dimension to Central Park's success.[92] The reworking of relations between nature and culture is pivotal to the sophistication of Central Park not only as an aesthetic advance on earlier developments in American landscape design but also as a measure of the park's skillful integration into the dynamics of real estate speculation and the commodification of the "first nature" of Manhattan Island. In this sense the park represents a kind of elaborate spatial fix to the economic downturn of the 1850s and was a precursor to changing patterns of investment that gathered pace from the 1880s onward in a new phase of global capital accumulation within which New York was to play a growing role. The speculative dimension to the creation of Central Park introduces the pivotal contribution of abstract space to the commodification of nature, in which capital becomes aestheticized as an imaginary fragment of "first nature" replete with lakes, meadows, and vine-covered rocks. Yet in Central Park we fail to find a clear Lefebvrian transition between absolute, historical and abstract space but rather an alternative periodization superimposed in its place: the pretext for the park's creation was born out of the abstract commodification of nature, yet the subsequent use of the park has created a kind of imaginary "absolute space" in the center of Manhattan Island rooted in a primal mythology of urban origins.

2.5 Emerald Dreams

Central Park has been a recurring focus of interest for contemporary landscape architects, planners, and artists who seek to explore a more productive relation between nature and culture than that provided by the twentieth-century legacy of technological modernism. The park has become a powerful symbol for a kind of harmonic mediation between society and nature that allows the semblance of organic continuity to an imaginary American *Gemeinschaft* embodied in the Jeffersonian ideal. In contrast to this widely held position, this chapter has argued that the park is best conceived as a landscape that exemplifies some of the most powerful aesthetic and ideological dimensions to the dynamics of nineteenth-century capitalist urbanization. It is a landscape that demonstrates a highly sophisticated approach to the commodification of nature, not its antithesis as has so often been erroneously suggested. The Dutch architect Rem Koolhaas, for example, in an otherwise perceptive essay, suggests that "Central Park is not only the major recreational facility of Manhattan but also the record of its progress: a taxidermic preservation of nature that exhibits forever the drama of culture outdistancing nature."[93] In reality of course Central Park is not a preservation of nature but an example of "*seeming* Nature" (to borrow Henry Cleaveland's phrase) produced in accordance with the particular aesthetic predilections of mid-nineteenth-century cultural elites. Although the greensward plan for Central Park appears radically different from the geometrical abstraction of the earlier gridiron plan for the city, both landscapes reflected the outlook of the city's political and cultural elites: the decisive difference between the two lies in the aesthetic and ideological mediation of the relationship between nature and culture. While the initial city grid rested on a Promethean obliteration of the "first nature" of Manhattan Island, the Arcadian vision of Central Park exulted in an imaginary nature that in no way contradicted the fundamental dynamics of nineteenth-century urbanization. Perhaps it would be more useful to reframe our understanding of this relationship in terms of the social production of nature under modernity, with Central Park representing a resolution to this tension in the specific context of nineteenth-century American urbanization. The incorporation of a water reservoir into the fabric of the park's design reveals a far-reaching utilitarian and aesthetic synergy between the new park and the reshaping of relations between urban society and metropolitan nature. Both the creation of Central Park and the earlier construction of the Croton Aqueduct represent a significant realignment in the relationship between nature, capital, and urban space. They represent, above all, the reworking of the raw materials of nature into a new synthesis with modern society, in order to harvest the aesthetic and biophysical properties of the natural world to human advantage.

The achievements of nineteenth-century urban planning, both real and imagined, continue to cast a powerful hold over contemporary thinking on cities and urban design. Nature-based designs have been a defining element in virtually all conceptions of the urban ideal from the garden cities of Ebenezer Howard and Patrick Geddes to *La ville radieuse* of Le Corbusier and *The Disappearing City* of Frank Lloyd Wright. Just as nineteenth-century urban planners and social reformers repeatedly drew on nature as a means to create a more manageable and humane kind of urban society, we find similar sentiments today emerging from the ashes of the discredited modernist attempts to control urban space. The architect Peter Calthorpe, for example, suggests that "nature should provide the order and underlying structure of the metropolis."[94] In a similar vein, the architectural critic Charles Jencks concludes his analysis of the Los Angeles riots by invoking an urban bioregionalism drawing heavily on ecological analogies of diversity and interdependence.[95] Yet such perspectives invariably treat the city as a discrete sociospatial unit unrelated to any broader social and economic processes. These ecological conceptions of urban form believe that "nature," however ill-defined, provides a blueprint capable of contributing meaningfully to the advancement of city planning and social policy.

Olmsted's conception of the role of nature within urban design was born out of a preservationist instinct to protect (or recreate) natural fragments within an unchallenged urban whole. His role in the creation of the first wilderness park in the Yosemite Valley during the 1860s and his opposition to the destruction of the Niagara Falls by water power generators in the 1880s exemplify his approach. The protection of fragments of wild nature not only marks a distinctively American contribution to the emergence of environmental ethics but also provides a link between the Ramble of Central Park and the wider development of an American philosophy of nature that combined urban aesthetic sensibilities with an increasingly sophisticated scenic panorama. During the twentieth century Olmsted's ideas were developed further by Lewis Mumford, Philip Lewis, Ian

McHarg, and a generation of radical urban scholars, landscape designers, and architects, but the underlying faith in professional and technical elites was ultimately to undermine the wider legitimacy of environmental planning from the 1960s onward (a theme we will explore in subsequent chapters).

Central Park was the first large-scale meticulously planned urban park in America. Its significance contributed to the emergence of both urban planning and landscape architecture in America as powerful and respected professions with legitimate roles in urban government. Olmsted himself was to play a key role in designing the parks of Buffalo, Chicago, Montreal, Detroit, Boston, Rochester, Louisville, and many other cities across North America. The park-building program extended into the Progressive era as reformers sought to promote the health and morality of working-class citizens. In subsequent decades, the United States Parks Service, the Environmental Protection Agency, and a variety of other federal interventions in environmental management emerged from the expanded role of the state which grew out of nineteenth-century concerns with public health, social order, and advancement of the public good. Yet all these developments have stood in a precarious relation to the exigencies of capital accumulation, real estate speculation, and the counterdiscourses of laissez-faire public policy making. A central paradox running through Olmsted's contribution to the emergence of urban planning derives from conflicting conceptions of order in urban space: on the one hand, a series of conceptions of a natural order embracing an urban pastoral vision of social stability and aesthetic harmony; on the other, a conception of urban space as a rational grid or network amenable to management and control. It is ironic, therefore, that in the late 1860s Olmsted and Vaux designed the first multilane landscaped highway (Eastern Parkway, Brooklyn) and the first garden suburb in North America (Riverside, Illinois). These new developments would in time foster a countervision of "autopia," to use Reyner Banham's term, which would become the focus of a ferocious critique developed by Olmsted's successors in the organic wing of twentieth-century urban planning.[96] The critical point is that Olmsted and his future champions never understood the dynamics of capitalist urbanization. Whereas Central Park had been created in response to the physical and social chaos of rapid urbanization in nineteenth-century America, the development of new transport technologies enabled urban elites to escape the confines of maladministered urban space, leading toward an urban landscape of ever greater separation (see chapter 3). The full implications of this disjuncture were not to become apparent until the rapid social and spatial polarization of the 1960s, but its roots stem from the inherited political and design legacy of nineteenth-century urbanism.

Central Park has succeeded as a public space in spite of the ambiguities behind its creation. From the outset, the "public" arguments for a park emphasized a search for order, a harmonious balance between nature and culture rooted in organic analogies of a healthy city as a means to facilitate greater economic prosperity and ensure social harmony. In reality, however, it was sophisticated and farsighted economic calculations on the part of powerful merchants, land speculators, and property owners that carried the day. Olmsted's aesthetic vision proved acceptable to the city's political and economic elite because it powerfully inflated land values at a crucial juncture in the city's metamorphosis into a world city. The fact that Central Park has been admired and appreciated by generations of New Yorkers is an ironic outcome of the combination of an Anglophile aesthetic vision with sophisticated real estate speculation. The transformation of Central Park into a popular and enduring public space disrupted Olmsted's rarefied vision yet reveals the extent to which the park was as much the creation of a whole city and its people as the work of any single individual. In the final analysis it is the richness of social life—and its role in the material transformation of space—that dispels mythic conceptions of urban design and most effectively challenges those sectional interests that masquerade under the cloak of an imaginary public weal.

3

Technological Modernism and the Urban Parkway

> The traffic multiplied, concrete lanes moving laterally across the landscape.
>
> —*J. G. Ballard*[1]

> Elevated sidewalks, Gargantuan skyscrapers that would make the pyramids of Egypt picayune beside their colossal bulk, bridges connecting their cloud-piercing tops, airplane landings 500 feet above a seething man-made canyon—America materialized in an age of Steel—man's nearest approach to nature's most striking work of architecture, a mountain range!
>
> —*Hamilton Wright*[2]

From the 1920s and 1930s onward, the nineteenth-century infrastructural shell of New York City was gradually encased within an expanding network of new roads and bridges. Like fungal hyphae engaged in a saprophytic feast, these new structures hollowed out parts of the old city and produced a new urban landscape. The nineteenth-century legacies of Beaux-Arts eclecticism, the city beautiful traditions, and the Olmstedian picturesque were gradually displaced by a larger-scale urban vision with a more ambitious sense of interaction between the social, functional, and aesthetic dimensions to landscape design.[3] For the design historian Klaus-Jürgen Sembach, the period 1927 to 1934 marks the emergence of a new kind of international modernist sensibility crystallized in a state of dynamic tension with the political and economic turmoil of the time.[4] Within this aesthetically defined epoch, the great crash of 1929 stands out as a critical dividing line between past and future, a point where existing conceptions of the relationship between state, economy, and society inherited from the nineteenth century were dispelled. The Fordist New Deal era is widely considered to be the high point of technological modernism in the United States. From the architecture of Albert Kahn to the murals of Diego Rivera, the machine emerged as the preeminent motif for both industrial production and social organization.[5] Yet at the same time, particularly in the wake of World War I, a counterdiscourse of fear and anxiety surrounding technology also began to emerge. Critiques of technology developed by Max Weber, Edmund Husserl, and José Ortega y Gasset, for example, emphasized the existential dilemmas posed by the increasing intrusion of a technical rationality into the lived spaces of everyday life.[6] Other concerns focused on the kind of organizational and institutional mechanisms needed to mediate the impact of technological change on modern societies. In terms of urban planning, however, there was a widespread embrace of new technological applications to the management and design of cities. The counterdiscourses of suspicion toward technology would, for several decades, be subservient to the modernizing zeal of New York's political and economic elites; but in the post-World War II era the fault lines of urban design would become gradually exposed to ever wider scrutiny.

The interwar era marks the emergence of technology as a defining element in twentieth-century history. This is not to argue for some kind of technological determinism but rather to recognize how we can read different elements within the urban environment as cultural artifacts rich with the complexity of their own time and place. Consider, for example, Mies van der Rohe's German Pavilion at the International Exposition in Barcelona (1929); Le Corbusier's Pavillon de l'Esprit Nouveau at the Paris Exposition des Arts Décoratifs of 1925; the Milan Triennale exhibitions; or the Los Angeles architecture of Richard Neutra. All these tell a specific story within the broader narrative of an emerging modernist aesthetic and new conceptions of urban design. By the 1930s the modern movement had taken hold of the American imagination, marked by developments including the influential exhibition "Modern Architecture" held at the Museum of Modern Art in 1932 and the arrival of émigré architects such as Walter Gropius and Ludwig Mies van der Rohe.[7] Leading modernist planners and architects such as Frank Lloyd Wright, Thomas Church, Garrett Eckbo, and Jens Jensen sought to create a new synthesis between society and nature in the regional landscape. The regional vision of Wright, for example, sought to mold urban form to the materials and characteristics of unique places and dispense with the formalist and neoclassical motifs inherited from the nineteenth century.[8] The impact of modernist aesthetics in the New York metropolitan region became associated with large-scale infrastructure projects such as roads and bridges; notably with the interwar development of landscaped roads, known as "urban parkways," which contributed to a new synthesis of nature, technology, and landscape design.[9]

The individual most closely associated with this urban transformation is Robert Moses, who headed a variety of park and planning agencies for New York City from the late 1920s till the early 1960s. For Marshall Berman, in his definitive study of twentieth-century modernity, Moses stands out as the key figure who "was able to preempt the vision of the modern" for forty years.[10] In a similar vein, Blair Ruble notes that Moses had "more power to shape the metropolis than any individual ever had before or since."[11] Initially, the new multilane highways he created were seen as a marvel of the modern metropolis, combining unprecedented mobility with completely new urban vistas. When the West Side Improvement in Manhattan was opened in 1937, for example, the *New York Times* described it as "one of the most magnificent urban highways on earth." Yet by 1973 the *Times* was referring to the same highway as "an ugly traffic wall between the city and the river."[12] The eventual exhaustion of the Moses era in the 1960s marks the end of a long chapter in urban history and is inextricably linked with the dissipation of technological modernism as a progressive force for social change and the decline of the "urban totality" as a meaningful focus for American planners.

3.1 The Automobilization of the American Landscape

> The remodeling of the earth and its cities is still only at a germinal stage: only in isolated works of technics, like a power dam or great highway, does one begin to feel the thrust and sweep of the new creative imagination: but plainly, the day of passive acquiescence to the given environment, the day of sleepy oblivion to this source of life and culture, is drawing to an end.
>
> —*Lewis Mumford*[13]

In the 1920s and 1930s the United States emerged as a pioneer in the construction of landscaped roads or "parkways." These new features of the American landscape form part of a steady progression from the locally based urban beautification phase of tree-lined boulevards, to the regional extension of urban roads into sparsely developed fragments of nature at the edge of cities, and finally, to the development of a national network of federally promoted recreational routes.[14] The first urban parkways in North America were the Eastern and Ocean parkways built in Brooklyn during the early 1870s, which were inspired by the tree-lined boulevards of nineteenth-century Paris and Berlin.[15] The innovative designs developed by Frederick Law Olmsted, Calvert Vaux, and Charles Eliot in late nineteenth-century America were to form the basis for the urban parkways of the twentieth century with their emphasis on uninterrupted travel along landscaped roads.[16] The early parkways were promoted by the city beautiful movement to provide landscaped arterial routes in the face of urban congestion and uncontrolled development. But their origins in Brooklyn, Boston, Cleveland, and elsewhere also marked the development of a new kind of "metropolitan park" devoted to the latest forms of leisure and mobility.

The first of a new generation of modern arterial parkways specifically designed for use by automobiles was the fifteen-mile Bronx River Parkway, completed in 1925, which extended from the northern part of the city into Westchester County. Originally conceived by Frederick Law Olmsted in the 1890s, this stretch of landscaped road was aimed at "pleasure type" vehicles and employed curved designs to emulate the beauty of the landscape through which the road would pass (figure 3.1). The road was conceived as part of a wider project to transform the regional landscape and provide the city with a "magnificent approach" to the newly completed Kensico Dam as a means to "inspire civic pride in the citizens of New York."[17] The final report of the Bronx Parkway Commission reflected on the scale of the transformation along the new road:

> The same observer, had he walked up the valley of the Bronx River from Bronx Park to Valhalla in 1913, would have seen many buildings ranging from mere shacks to substantial factories, stretches of waste swamp and dismal land stripped of all natural beauty, and a foul, ill-smelling and uncontrolled stream spreading its polluted waters and refuse upon the adjacent lowlands . . . unsightly poles and electric wires, hideous billboards aggregating seven miles in length and many other visible evidences of the destruction of natural beauties and the waste and ugliness all too common in the less favored portions of suburban towns.[18]

For the project's principal landscape architect Hermann W. Merkel, the completion of the Bronx River Parkway marked "a distinct epoch in municipal and suburban park work." Just as Central Park had been an imaginary landscape created in the heart of the city, the Bronx River Parkway was to be a scenic utopia reflecting an "authentic" American landscape. The parkway verges were planted with thirty thousand trees, including only oaks, maples, pines, and other trees and shrubs which "were or might have been indigenous" to the local area.[19] Far from destroying the "natural" landscape, then, these early parkways were conceived

3.1 The Bronx River Parkway (1922).
Source: Courtesy of the Westchester County Archives.

as part of a larger rationale of landscape restoration to create "an environment of unsurpassed beauty."[20]

The new parkway thus reflected the landscape ethos of the Progressive era with its desire to foster a new sensitivity toward the ecological heritage of America's past, but it also signaled a more radical form of regional organicism that would find its ultimate expression in the landscape design of Garrett Eckbo and Frank Lloyd Wright. Wright, for example, welcomed the automobile as a means to foster a new kind of decentralized urban form and a closer engagement with the natural world.[21] In seeking to screen out rural slums, the parkway design masked its own origins as simply another facet of the development process that had begun to transform the landscape of the urban fringe.[22] Though the construction of the parkway had been vigorously opposed on grounds of cost, it soon yielded a spectacular increase in the value of adjacent land and property: by the early 1930s tax revenues for Westchester County far exceeded what might have been expected with no improvements in the local landscape and its arterial roads.[23] Yet the single-minded purpose behind parkway design, however aesthetically pleasing, would gradually reveal an inner tension between the democratic ideals of American modernism and the technocratic logic of landscape change. And the nativist doctrine of landscape authenticity belied deeper fears about the aesthetic and ecological contamination of the American countryside, which would reach their ultimate expression in the interwar German *Autobahn* designed by Alwin Seifert.[24]

The Bronx River Parkway's most significant design feature was that it ran continuously from New York City into the surrounding countryside without any interruptions from cross traffic. This opportunity for continuous movement facilitated easy access to and departure from the city and contributed to the emergence of suburbia as an increasingly dominant urban form.[25] It is especially significant, however, that this early generation of landscaped parkways, both in Europe and North America, was not constructed in order to meet the pressing demands of urban transportation but to provide an opportunity to escape from the congestion of the nineteenth-century city. One of the landscape architects for the Bronx River Parkway, Gilmore D. Clarke, observed at the height of the postwar highway construction program in the United States: "Control of access is at the very heart of the highway problem today. The general public has not become fully aware of the process by which *unlimited*-access highways lose their effectiveness and become obsolete for the purpose and the traffic volumes they were designed to serve."[26]

A distinctive feature of the new generation of parkways was the meticulous attention to landscape design and the masking of any apparent discontinuity between the natural and the artificial. Clarke's design for the Taconic State Parkway, for example, provides a vivid example of a new kind of mediation among nature, technology, and society, with what appears to be a delicate balance between the new infrastructural project and an imaginary natural order. Implicit within this aesthetic dialectic is the notion of engineering as an art form that can in some way embellish or even improve upon nature: there is no radical disjuncture here but a sense of aesthetic progression and purity of form. The new parkway or highway was to be emblematic of a combination of engineering science with the aesthetic sophistication of landscape architecture in the service of a "modern public." For Eckbo, the Taconic Parkway stands out as "one of the earliest and best of our truly scenic parkways," providing "as lovely an integration of highway engineering and landscape architecture as one could hope to find."[27] Sigfried Giedion, writing in 1941, described how the American parkway was "born out of the vision of our period . . . laid into the countryside, grooved into it between gentle green slopes blending so naturally into the contiguous land that the eye cannot distinguish between what is nature and what [is] the contribution of the landscape architect."[28]

The urban parkway represented a new spatial configuration of society, technology, and nature. In the 1920s the role of landscaped roads as an opportunity for new forms of leisure and visual pleasure gathered pace with the completion of the Arroyo Seco Parkway in southern California and the cross-country Lincoln Highway extending all the way from New York City to San Francisco.[29] Yet these new means of mobility would soon transcend their earlier association with leisure and become an integral component of the new urban landscape. Nature became simultaneously more distant (framed by the window of a moving car), more accessible (through greater public contact with remote areas), and at the same time more individualized as an aesthetic experience. Just as the nineteenth-century extension of railways generated new forms of "nature tourism," so the spread of car ownership fostered new types of cultural engagement with nature. Nature was now a panoramic experience, the tactile and olfactory senses subsumed by an emphasis on separation, movement, and visual power:

> As with many other creations born out of the spirit of this age, the meaning and beauty of the parkway cannot be grasped from a single point of observation, as was possible when from a window of the château of Versailles the whole expanse of nature could be embraced in one view. It can be revealed only by movement. . . . The space-time feeling of our period can seldom be felt so keenly as when driving, the wheel under one's hand, up and down hills, beneath overpasses, up ramps, and over giant bridges.[30]

Between 1923 and 1933 the accessibility of New York City was gradually transformed by the building of nearly 100 miles of parkways and freeways, including the Saw Mill River, the Hutchinson River, the Briarcliff-Peekskill, and the Cross County parkways. Yet these new roads quickly proved inadequate for growing volumes of traffic. The first generation of urban parkways were conceived in an era before the rise of road haulage and chronic traffic congestion: early plans and sketches depict virtually empty roads in which driving is portrayed as a new form of leisure.[31] As late as 1900 there were only 8,000 vehicle registrations across the United States. By 1916 this figure had leapt to nearly four million, rising rapidly to over twenty million by 1925. In New York City cars first outnumbered horses as early as 1917 and soon became the dominant means of private transportation. Between 1920 and 1926 the total number of registered motor vehicles in the New York metropolitan region grew from 540,000 to 1.3 million. Rising levels of car ownership in the 1920s proved central to the emergence of Manhattan as the "epicenter of a new culture of consumption."[32] The automobile became a formative aspect of newly emerging patterns of production, consumption, and leisure. By the mid-1920s the production of cars had become the most significant sector of American industry as measured by product value. Car production proved so central to the development of twentieth-century American society that this historical epoch, from the interwar period until at least the late 1960s, has been widely termed "Fordist" (or in some cases "Sloanist" after Alfred P. Sloan, then head of General Motors) to indicate the particular conjunction of social, cultural, and organizational forms that underpinned this era of rapid and sustained economic expansion and the spread of new consumption norms.

These new approaches to the design of urban space were not restricted to the United States: we can find similar developments elsewhere in the rapid expansion of the German *Autobahn* and Italian *autostrada*. In Martin Wagner's plans for 1920s Berlin, for example, the needs of the motorcar were paramount in combination with the development of new peripheral housing estates. Similarly, in Fritz Schumacher's plans for Hamburg (1909) and Cologne (1920) the centers of these cities were opened out with parks and public spaces. By the 1930s, however, the earlier modernist conceptions of urban space were gradually displaced by new conjunctions of technology, power, and urban form. In Italy, the modernist ideals of Le Corbusier were transformed into the fascist architecture of Giuseppe Terragni, Luigi Figini, and Gino Pollini; in the Soviet Union, tensions opened up between the traditional approaches of Goltz and Shchusev and those of the constructivists (Krutikov, Lavrov, and Popov); and in Germany, earlier modernist approaches were supplanted by the neoclassical monumentalism of Speer.[33] What is of interest, however, is the peculiar malleability of different conjunctions of technology and urban form, within which the emphasis on speed and mobility emerges as a defining element in twentieth-century modernity: "From Marinetti through the European embrace of Taylorism, to the productivism of the Vkhutemas and the Bauhaus, speed, whether dynamizing aesthetic perception, augmenting production, or boosting consumption appears as the ubiquitous common term linking modernity to diverse instances of twentieth-century modernism."[34]

We can trace the origins of new thinking about cities, technology, and mobility to the post-Haussmann era of land use zoning and regional planning which developed across much of Europe during the 1880s. The Haussmann approach toward comprehensive urban reconstruction was to be emulated by Joseph

Stübben in Cologne, Ildefonso Cerdà in Barcelona, H. P. Berlage in Amsterdam, and Otto Wagner in Vienna. In the early twentieth century the segregated and hierarchical ordering of urban space was further advanced by the French planners Eugène Hénard and Tony Garnier. Progressively greater emphasis was to be placed on the radical separation of land uses and the creation of the "hygienic city" in which light, air, and movement would take precedence, ending the congested mingling of land uses in the old city centers. Hénard, for example, was an early advocate of the "death of the street," which was to be replaced by a four-tiered series of concourses, part underground and part overground, separating different kinds of traffic.[35] These new conceptions of urban infrastructure were modified yet again by the Italian futurist Antonio Sant'Elia with his emphasis on multilevel roadways as a means to perfect the circulatory dynamics of urban space.[36]

By the late 1920s we can identify a distinctively American variant on the avant-garde city marked by the increasing scale of the urban imagination applied to the reshaping of the modern metropolis. The congested streets of Manhattan became the focus of a succession of futuristic designs. In 1924, for example, the architect Harvey Wiley Corbett described the need for a vertical separation of different flows of "foot, wheel and rail" to create a multidecked urban landscape of speed and mobility freed from the "congestion and turmoil of the present."[37] Just three years later, the leading transport planner John A. Harriss, appointed by Mayor Walker in the pre-Moses era, was calling for a six-deck multilane highway system connecting New York City to every other East Coast city. Harriss envisioned the "American Multiple Highway" as a means to solve Manhattan's traffic problems for 100 years using "a network of weather-protected self-ventilating highways."[38] And in *La ville radieuse,* published in 1933, Le Corbusier amplified these themes by calling for the replacement of nineteenth-century New York with a high-speed modern metropolis.[39] Le Corbusier argued that the city had to adapt to the technological logic of capitalist urbanization. The extending network of urban parkways was envisaged as a triumph of "scientific management," rooted in the rational ethos of technical engineering.

By the mid-1930s the earlier generation of New York parkways had become outmoded through a combination of increased use and improvements in automobile design. The stage was now set for a far more extensive transformation of the urban landscape founded on a fusion of centrally directed national purpose in combination with the immense power of the automobile within American economy and society. At the 1939 New York World's Fair the Futurama pavilion caught the attention of the city planner Robert Moses, who was rapidly emerging as a key player in the rebuilding of New York City. Of particular interest to Moses was a model city project designed by the American architect Norman Bel Geddes and commissioned by the Shell Oil Company and General Motors.[40] This vast installation depicted a structural transformation in the physical fabric of cities which marked the ascendancy of the car over all other means of urban transport.

3.2 Robert Moses and the Radiant City

> Road building must be viewed in an entirely different light than it has up to now. It has to be considered as something far more than merely providing the means for getting people from one place to the next. The motorways must be considered as an essential part of the entire economic system of the country. . . . As the American road builder becomes a planner, he will grow into a key individual who is responsible to the whole nation.
>
> —*Norman Bel Geddes*[41]

> If Le Corbusier was just theorizing about a city in which to enjoy nature, the New York of Robert Moses offered its citizens, with the automobile, the chance to go out and get it for themselves.
>
> —*Christian Zapatka*[42]

The reconstruction of the New York metropolitan region in the New Deal era has come to be closely associated with the forty-year career of Robert Moses (1888–1981), extending from his appointment as president of the New York

State Parks Council and the Long Island State Park Commission in 1924 through to his somewhat ignominious last public office as president of the 1964–1965 New York World's Fair. In 1934 the newly elected Mayor Fiorello H. La Guardia consolidated Moses's growing power by appointing him as the sole commissioner of a unified and centralized Department of Parks for the whole city. Moses headed seven separate public agencies concerned with parks and worked assiduously to build up "an intricate network of interlocking commissions and public authorities."[43] One of his own executive officers noted in 1938, for example, that it is difficult "for an outsider to tell where the jurisdiction of one organization under Park Commissioner Moses ends and another begins."[44] Moses owed his power to the unique conjunction of political and economic forces that constituted the American New Deal in combination with the modernizing zeal of the regional business interests that dominated emerging conceptions of urban planning and shared his vision for an integrated, car-oriented urban form for the New York metropolitan region (figure 3.2). His political skill enabled the realization of major urban reconstruction, while his rivals' plans languished as no more than sketches and drawings in newspapers and professional journals.

After the failure of private enterprise to secure economic prosperity during the Great Depression, there was a fundamental realignment of public policy under the New Deal that sustained a broadly consensual Keynesian pattern of policy making from the early 1930s to the mid-1960s. During the 1930s large-scale construction activity was transformed from a private to a public enterprise, facilitating a new approach to comprehensive urban reconstruction. Virtually every large project that was built in the 1930s was built with federal money under the auspices of the great New Deal agencies. Parks and highways were to have the highest share of this New Deal spending. The new construction projects were planned around a series of complex and well-articulated social goals: the stimulation of economic activity; the need to put millions back to work; the desire for greater social harmony; the modernization of regional economies; the utilization of new technologies; and above all, the enlargement of "the meaning of 'the public,'" in order to "give symbolic demonstrations of how American life could be enriched both materially and spiritually through the medium of public works."[45] Yet as we

3.2 Robert Moses in *Fortune* magazine, June 1938.
Source: From the collection of the MTA Bridges and Tunnels Special Archive.

shall see, the popular association of the Moses era with both "public space" and the wider ideals of the New Deal is far from straightforward.

The first significant achievement of Robert Moses, as Long Island State Park Commissioner, was the completion of Jones Beach State Park in 1929. Jones Beach, the first of a radically new generation of mass recreational facilities for the city, was built with parking lots for thousands of cars—an unprecedented feature. The new roads allowed affluent New Yorkers the chance to escape the city, the roads "carefully twisting and turning to avoid unnecessary desecration of the North Shore's carefully manicured estates and golf courses."[46] For Marshall Berman, this glittering network of parkways and beaches was a new dimension to the "modern pastoral," offering "a spectacular display of the primary forms of nature—earth, sun, water, sky—but nature here appears with an abstract horizontal purity and a luminous clarity that only culture can create."[47] Berman's description of Jones Beach captures something of the undoubted critical acclaim and public delight in these innovative architectural forms that gave physical expression to the changing aspirations and aesthetic sensibilities of interwar American society. Contemporary commentators such as Lewis Mumford were no less effusive: "The landscaping of the highway and the beach leaves me in a state of ecstatic admiration. . . . The word 'heavenly' should be reserved for such complete achievements."[48] Though Moses was not trained in any formal aspect of planning, architecture, or engineering, he drew on some of the most remarkable talents of the time, such as Othmar Amman, Aymar Embury II, Gilmore D. Clarke, and Herbert Magoon, in order to strive toward his vision of "urban perfection."[49] Moses saw the conception of new parkways and recreational facilities as a continuation rather than a radical break from earlier conceptions of urban form. Indeed, many of the major schemes of the Moses era—the Henry Hudson Bridge, Grand Central Parkway in Queens, and the Brooklyn circumferential parkway—had all been conceived in previous years. The circumferential road system had been advocated by the Regional Plan Association as early as 1929, while other projects had been anticipated in Edward H. Bennett's plan for Brooklyn in 1914. Of increasing significance was the role of parkways as conduits for rapidly growing commuter traffic in the 1920s and 1930s, which gave a sense of urgency to Moses's scheme for a "circumferential parkway" encircling the entire city. Growing volumes of motorized traffic from Connecticut, New Jersey, and upstate New York had to pass through the congested streets of Manhattan to reach destinations beyond the city, and most of the existing bridges had been designed for horses rather than cars and trucks.[50]

The New York Regional Plan, set up in 1921 by the Russell Sage Foundation, quickly emerged as the pivotal strategic forum for responding to the challenges of urban change. The plan brought together a diverse array of social scientists, planners, urbanists, architects, and engineers. By the early decades of the twentieth century, as we saw in chapter 1, the engineering profession was imbued with a sense of omnipotence in the management of large metropolitan areas. Contemporary writings suggest a yearning for a more active role in urban governance by New York's planners and engineers, eager to transform the urban environment and create a "Wonder City of the World."[51] The Regional Plan transformed the earlier planning ideals of Lewis, Olmsted, and Burnham "with additional circumferential highways and radial routes in the hinterland, double- and triple-decker traffic flows within the center, and new bridges linking Manhattan, Queens, and the Bronx."[52] These new dimensions to urban infrastructure were integral to a broader dynamic that would irrevocably alter both the form and function of urban space: "If the late nineteenth century parkway plans of Olmsted, Vaux and Eliot implied a tendency toward suburbanization, this tendency gradually became institutionalized with the parkway plans of the 1920s and 1930s, which provided successively more expedient access in and out of the city until they assumed the role of expressways."[53]

One of the key developments that facilitated the gradual integration of the New York metropolitan region was the innovative Triborough Bridge project. Opening in 1936, this unique structure consisted of three steel bridges linking the boroughs of Manhattan, the Bronx, and Queens for the first time (figure 3.3). By a curious irony, construction commenced on 25 October 1929, the same day as the stock market crash. As a consequence, construction activity was quickly halted when investors became unwilling to purchase municipal bonds. In re-

3.3 Berenice Abbot, *Triborough Bridge: Steel Girders* (1937).
Source: Courtesy of the Museum of the City of New York. Federal Arts Project: Changing New York.

sponse to this changed set of circumstances, the Triborough Bridge and Tunnel Authority was created in 1933 as an alternative means to finance large-scale urban infrastructure. The new authority quickly emerged as a focal point for Moses's construction activity through its collection of tolls, which allowed a degree of independence from state and federal sources of funding: in its first year of operation, for example, the bridge carried over nine million vehicles and generated nearly $3 million in tolls.[54] The Triborough Bridge project also involved the breaking of Tammany control of public works in the city with the election of the La Guardia fusion administration in 1933. This decisive political shift brought about a radical expansion in the New Deal emphasis on public construction projects and further advanced the design agenda of Moses and the Regional Plan Association.

The momentum behind urban reconstruction began to gather pace as Moses consolidated his political power base within the city and secured new sources of funding. The combination of tolls and new sources of public money provided the means to realize a succession of new projects. In 1938, for example, Moses made a direct appeal to the New York State Legislature:

> The need for providing funds for the construction of these important arterial parkway connections can hardly be overestimated. Landscaped parkways with restricted frontage, and without crossings at grade, or traffic lights, confined to pleasure vehicles, have proven to be the most efficient way of providing for a smooth flow of traffic. Unfortunately most of the new parkways . . . do not penetrate the city much beyond its border. *It is now proposed to extend them into the heart of the city* and to eliminate the bottlenecks in the Cross County Parkway just north of the city limits so that the whole network of parkways in the metropolitan area can be tied together into one unified system.[55]

In 1944 the city's circumferential parkway was finally completed, with over one hundred miles of parkways and express highways. One of its most spectacular pieces was the so-called "pretzel" intersection between the Grand Central Parkway, the Grand Central Parkway Extension, and the Interborough Parkway to produce a multilevel road with five underpasses. For Sigfried Giedion, the pretzel interchange was "one of the most elaborate and highly organized of all recent solutions of the problems of division and crossing of arterial traffic."[56] The new urban landscape of New York was a celebration of the latest advances in engineering science, the projects were spectacular in scale, and the overall impact attested to the primacy of mobility in urban design.

Yet this vast infrastructural legacy poses a series of methodological dilemmas for historical interpretation. Though Moses undoubtedly had a critical impact, not least through his immense legislative and managerial skill, various representations of him have systematically exaggerated his role.[57] The historical evidence clearly shows that Moses did not have unlimited power: he was prevented, for example, from constructing a Brooklyn-Battery bridge, from building expressways in Lower Manhattan, and from extending his planning powers to the city's airports.[58] The emerging primacy of the automobile was also underpinned by the failure of different railway companies to cooperate with attempts to coordinate alternative forms of transport, as envisioned in earlier incarnations of the New York Regional Plan. This failure to improve rail service ensured the long-term decline of the port of New York, contributed toward the devastation of the regional economy in future decades, and allowed the gradual ascendancy of financial over industrial capital in the region. The increasing predominance of the highway system facilitated regional decentralization and effectively ensured the predominance of the "suburban solution" to urban congestion.[59]

These developments were not restricted to New York: the interwar era saw comprehensive highway systems developed in Chicago, Minneapolis, Kansas City, Seattle, and elsewhere. The age of the automobile introduced a new kind of fluidity to urban form: the technical ethos of urban and regional planning was little more than a fantasy in the face of a dynamic of interrelated cultural, economic, and technological change that would transform the American landscape. Roads symbolized a kind of "structural continuity" in the evolving morphology of urban networks.[60] Their creation attested to the enduring dynamics of spatial organization and movement within an emerging regional industrial complex, but at

3.4 Artist's impression of the Highbridge Interchange, Cross Bronx Expressway (circa 1950). Source: Painting by Ted Kauteky. MTA Bridges and Tunnels Special Archive.

the same time their construction revealed a continuity in political and economic purpose that spanned the new constellations of state power and new forms of partnership with capital. In surveys of American road building from the 1940s, one finds little recognition of any contradiction between the dynamic of technological modernism and the lived space of cities:

> Highway design, in the broadest sense, rests upon landscape principles as well as upon the more commonly recognized engineering principles of alignment, profile, grade cross-section, roadway and right-of-way width, drainage, and structural strength and durability. . . . All these things may be done in complete consistency with the utilitarian functions of the expressways. And, so treated, these new arterial ways may be made—not the unsightly and obstructive gashes feared by some—but rather elongated parks bringing to the inner city a welcome addition of beauty, grace and green open space.[61]

Another difficulty in assessing Moses's impact has been the separation of aesthetic dimensions to his infrastructural projects from the wider social and economic dynamics of urban change underpinned by their construction. In particular, the relationships between the architectural innovation of his earlier achievements and the eventual dissipation of modernist planning ideals remains obscure. Moses developed a radical populist vision that was disdainful toward academic critics who "deliver homilies to long-haired listeners." "This town is too tough for them," mocked Moses, "and they had better keep out of the rough and tumble of the market place."[62] Given his outspoken antipathy toward a range of key figures associated with the modernist movement, such as Eliel Saarinen, Walter Gropius, and Erich Mendelsohn, it is difficult to conceive of Moses as a leading exponent of American modernism.[63] The scale of Moses's power has been tellingly compared with the vast infrastructural projects associated with interwar European fascism. In 1938, for example, an official visitor from Italy is reported to have been surprised that it was possible in a democracy to emulate "the swamp-filling, road-building efficiency and beauty of fascist public works."[64] These new features of the urban environment were met with a kind of strange fascination, as powerful public sculptures embodying the promise of a new technological age but also a sense of dislocation and foreboding.

A narrow focus on Moses's public works as artifacts in themselves overlooks the national and regional context within which this urban reconstruction took place. The real power of Moses and the Regional Plan Association was actually remarkably weak in relation to the underlying dynamics of urban change. Moses worked within the context of the Regional Plan, which in turn reflected the needs of New York's business elite. Their priorities were to focus business activity in lower Manhattan, redistribute urban growth outward, and restrict manufacturing industry to an intermediate belt within New Jersey and the outer boroughs.[65] These developments favored the continued dominance of lower Manhattan business interests over any alternative conceptions of urban form by endorsing the emerging role of the city as an international finance center serving a global economy rather than an integrated regional economy. By the early 1940s the construction of highways, parkways, and the new urban landscape of the Moses era had transformed New York from a national center into the international "capital of capitalism," reflecting the city's symbolic role at the center of a new world order.[66]

The association of the Moses era with the expansion of public space needs to be qualified by its emphasis on new forms of consumerist exclusivity and its willful disregard for the needs of the poorest urban communities. Behind "a persona of public-minded magnanimity" lies a far more complex relationship between Moses and the public he served.[67] The overriding social and spatial dynamic was one of separation rather than interaction. Moses largely disregarded black neighborhoods in the development of his parks program and obstructed or neglected the development of public transportation to new recreational facilities.[68] The parkways leading to Jones Beach, for example, "could only be experienced in cars: their underpasses were purposely built too low for buses to clear them, so that public transit could not bring masses of people out from the city to the beach. This was a distinctively techno-pastoral garden, open only to those who possessed the latest modern machines . . . a uniquely privatized form of public space."[69] Moses eschewed any clear connection with the theoretical underpinnings of professional planning, preferring instead to combine a populist urban vision with newly emerging middle-class aspirations. The urban transformation he favored was couched in the rhetoric of incremental change as the gradual accommodation to a new kind of American middle-class consumption. The new landscape was to be a "rational choice idiom" in asphalt and steel:

> This land of leisure and modernity had to be manifested for Moses, it appears, with expressions of accessibility, speed and the retrieval of open land, no longer for agriculture or wealthy seclusion, as had been customary in the past, but for the recreation of a new middle class. That new class, in exploring these frontiers, then became identified inextricably with its prize symbol of mobility and contemporaneity—the automobile.[70]

When combined with rising prosperity, higher levels of car ownership, and changing middle-class aspirations, the momentum of urban decline and sociospatial polarization fulfilled its own dynamic. Patterns of increased social and ethnic segregation and widening disparities in wealth emerged in virtually every American city during this period. The underlying dynamics of urban change were increasingly driven at a national level by federal policies toward the building of highways and housing, which provided huge public subsidies for the expansion of suburbs at the expense of the urban poor. As early as 1940, for example, the chairman of the New York City Planning Commission, Rexford G. Tugwell, warned that an "ugly and inhospitable" urban landscape was fostering demographic flight. "In the future," wrote Tugwell, "either we are going to have to bring the population back from the suburbs or we are going to allow the city itself to be ruined."[71] The prescient observations of Tugwell and other critics of American urbanism would only really take hold of the public imagination some twenty years later, when the planning debates of the 1960s began to expose the sharply different perspectives that had hitherto been subsumed under some imaginary conception of the public interest.

3.3 The Demise of Technological Modernism

> Perhaps our age will be known to the future historian as the age of the bulldozer and the exterminator; and in many parts of the country the building of a highway has about the same result upon vegetation and human structures as the passage of a tornado or the blast of an atomic bomb.
>
> —*Lewis Mumford*[72]

> You can draw any kind of picture you like on a clean slate and indulge your every whim in the wilderness in laying out a New Delhi, Canberra or Brasilia, but when you operate in an overbuilt metropolis you have to hack your way with a meat ax.
>
> —*Robert Moses*[73]

The critical moment when Moses's vision began to play a part in a far broader crisis in urban planning was his decision to extend major highways through heavily populated parts of the city. Though the principles behind this shift had been laid down with the completion of New York's circumferential parkway, it was his postwar decision to plough multilane highways through predominantly residential areas that precipitated a fundamental realignment in his relationship with progressive planners, architects, and the wider public he served. For admirers of Moses such as Giedion, the physical form of the modern city had become an obstacle to further progress: "The parkway ends today, and obviously must end, where the massive body of the city begins," wrote Giedion: "No facilities of approach can really accomplish anything unless the city changes its actual structure."[74] We can trace a theoretical and historical web here between the nineteenth-century transformation of Paris under Haussmann, the space-time architectonics of Giedion, and the ambition of Moses to transform New York City into a modern car-dominated metropolis.

During the 1960s the construction of urban parkways began to open up a conflict between the centralized engineering-dominated ethos behind infrastructural development and growing demands for greater public participation in urban policy making. Urban planning faced the disintegration of the kind of putative "public interest" that had sustained the ideal of comprehensive urban renewal through large-scale investments in infrastructure. Planners themselves increasingly recognized that "the ideal of master planning" was illusory and began to explore ways of bolstering their legitimacy through wider public consultation.[75] Yet American society was becoming more divided and fragmentary in the mid-1960s, which contributed to the growing marginalization of urban planning. In the wake of the Harlem and Los Angeles ghetto riots, planners could no longer conceive of the public interest as an unproblematic dimension to spatial and technological rationalization.

The construction of multilane highways through the heart of densely populated urban areas in the 1950s and 1960s eloquently reveals the weakness of community power in resisting the colossal momentum of the postwar highway construction program. From the late 1950s to the early 1960s, for example, the heart of the Bronx was destroyed to create the Cross Bronx Expressway, which carved a seven-mile swath through the East Tremont district (figure 3.5). The road "literally split the viable and healthy neighborhood of East Tremont in two, left its residents to contend with high levels of noise and air pollution, destroyed a large portion of its solid housing stock and retail space and, many believe, was the opening wedge for an urban rot that would make East Tremont, by the 1970s, a slum."[76] Patterns of infrastructural investment that had previously been conceived as integral to urban revitalization had now become directly implicated in postwar urban decline and the destruction of city life. It would be misleading, however, to claim that Moses was in some way solely responsible for the devastation of the Bronx, which was subject to a combination of pressures such as redlining, disinvestment, and industrial decline operating in a wider theater of urban change than any one individual could possibly control.

Moses responded to the mounting criticism of his work by emphasizing not only the economic necessity but also the aesthetic qualities of his infrastructural

3.5 The Cross Bronx Expressway under construction (circa 1956).
Source: MTA Bridges and Tunnels Special Archive.

projects. In 1966, for example, he argued that the beauty of public works surpassed that of nature: "Natural beauty we understand, but what is man-made beauty? Ornament is well enough in its place and the sculptor should work with the architect, but the greatest structural beauty man has created is the spiderlike, bare, unencrusted, unornamented suspension bridge, held aloft by cunningly woven wires, spanning the endless procession of ships carrying traffic from shore to shore."[77] Moses displayed a mix of disdain and incredulity toward his detractors, failing to recognize that the emerging challenge to urban planning in the 1960s was quite different in scale from the earlier criticisms of his work: "In pursuit of beauty and civilization, I have lived through not only violent criticism, which in the end usually proved groundless, but through almost incredible vilification and denunciation. Witness the opposition to every major enterprise my friends and associates have engaged in the field of public works."[78]

The politicization of highways policy in New York intensified during the late 1960s. In 1965 Robert Moses proposed that an elevated expressway be constructed through lower Manhattan, linking the Williamsburg and Manhattan Bridges on the east side and connecting with the Holland Tunnel to the west. First promoted by the Regional Plan Association in 1929, the so-called "Lomex" scheme would have involved building a six-lane highway on a 1.5-mile viaduct suspended above some of the most densely developed parts of the city (figure 3.6). In the 1969 mayoral election, John Lindsay campaigned for reelection on an anti-Lower Manhattan Expressway platform, but reversed his position after the election under pressure from powerful public agencies such as the Board of Estimate and business interests led by David Rockefeller. In response, the opposition to highway building became more sophisticated and better organized. Leading antimodern planners such as Jane Jacobs sought to raise the political profile of the campaign against the Lower Manhattan Expressway, in conjunction with the Sierra Club and a panoply of other newly emerging environmental groups. The road proposal became a powerful aspect of Jacobs's critique of postwar urbanism, which warned that such projects "eviscerate great cities." Jacobs became a leading critic of modern planning, comparable in terms of her impact with the

3.6 The city that was never built: the Lower Manhattan Expressway (elevated scheme through buildings at left) (circa early 1960s).
Source: Painting by Gero. MTA Bridges and Tunnels Special Archive.

contemporary critique of industrialized agriculture developed by Rachel Carson. The Lower Manhattan Expressway quickly became the most politicized of all the city's major road projects so far, its opponents raising issues of blight, dereliction, and air pollution. A November 1968 study predicted increased levels of carbon monoxide for residents in the vicinity of the proposed road. The project finally collapsed in 1971 after Governor Nelson Rockefeller withdrew his political support for the scheme.[79]

The growing sense of disquiet at this time over the reconstruction of postwar urban form was also shared by prominent modernist architects such as Lawrence Halprin:

> In the process of a singleminded approach to mobility, every other aspect of environmental design has been sacrificed, as though speed and mobility were the only and ultimate justification, with an overriding virtue of their own. As a result, freeways have cut great swaths through urban communities, whole neighborhoods have been sliced in half, parks have been segmented, waterfronts have been cut off from the body of the city, and the intricate, closely woven texture of the city's tapestry has been demolished.[80]

Given the agreement between Jacobs and Halprin on this point, the opposition to major road projects cannot be demarcated in terms of a simple modern versus antimodern divide. The critical problem was one of renegotiating the relationship between planning and urban design in a radically changed social and political arena. The 1960s marks a decisive turning point in the history of urban planning in New York. Moses found his power rapidly eroded: he resigned as parks commissioner in 1960 (in part to focus attention on the organization of the World's Fair) and was forced out of his remaining state appointments by Governor Rockefeller in 1962. In 1968 he left his last significant power base within the Triborough Bridge and Tunnel Authority, at a time of growing polarization in urban planning policy.

Despite the absence of Moses, the arguments for a comprehensive new highway system for Manhattan were to resurface again in December 1973 when a cement truck plunged through the dilapidated Westside Highway at Gansevoort Street, leading to an immediate closure of the entire road from West 46th Street down to the Battery. This gave sudden impetus to a project first mooted in the 1930s for a complete reconstruction of the transport infrastructure along Manhattan's west side. The original Westside Highway had been constructed between 1927 and 1931, running from 72nd Street south to Chambers Street. The road was extended between 1945 and 1948 to link with the Brooklyn Battery Tunnel, but by the late 1950s it was widely regarded as obsolete and inadequate for projected future increases in road traffic. Between 1957 and 1971 a variety of proposals were debated for the rebuilding of the Westside Highway. Finally, in March 1975 Governor Hugh Carey and Mayor Abraham Beame announced that they would support the "modified outboard" plan better known as the Westway scheme. In retrospect, the controversy over the proposed Westway project can be conceived as "the ultimate extrapolation of American highway policy."[81] This new road project was supported by capital (a wide array of property developers and business interests), organized labor (with the exception of one union, the Teamsters local 703, representing the fishing industry), and by every tier of government from City Hall to the White House. The city's fiscal crisis of the 1970s added a sense of urgency to the project as a means to bolster New York's economy in hard times. In 1976, for example, the manager of the General Contractors Association of New York, William C. Finneran, Jr., declared that "to eliminate Westway, as it is now planned, would not only be a tragic loss to the economy of New York City but another unconscionable blow to an industry which has been the greatest victim of the fiscal crisis in the past, and has been one of the greatest contributors to the New York City economy."[82] In the mid-1970s Westway appeared inevitable; but its supporters underestimated the scale and complexity of the opposition.

The collapse of the consensus over highway construction in the 1960s and 1970s closely mirrors the broader dissolution of the New Deal bipartisan consensus in public policy. In New York, the emergence of well-organized opposition to road building can be traced to the coalition of small businesses and homeowners that defeated plans for the Lower Manhattan Expressway in the 1960s. Yet the opposition to Westway went far beyond these earlier protests to encompass a generalized indictment of the expansion of car ownership, consumerism, materialism, militarism, and the perceived misuse of public money. The New Deal consensus sustained by inexorable economic growth was now being directly challenged as an irrational basis for public policy. Westway embodied the inherently contradictory impulses toward rational order and personal freedom that underpinned the postwar economic and technological transformation of urban life but which had exploded into open conflict by the late 1960s. "Westway is a metaphor for the city," declared Jane Jacobs, "how one feels about it depends on one's vision of what the city is and what it can become." In particular, Jacobs warned that the road would "Los Angeles-ize New York" and create a city "built for automobiles and not for people."[83] The affluent and culturally diverse neighborhoods of Greenwich Village and Chelsea gathered an increasingly elaborate and vocal band of supporters, drawing together expertise from the media, law, ecological science, and numerous other fields to reach a new level of sophistication in grassroots political lobbying.

The project was repeatedly lampooned as a "million-dollar boondoggle" that had subverted the purpose and resources of public policy. Opponents of the Westway project were adept at exploiting a 1973 decision by Congress to allow municipal governments to trade money committed to unbuilt sections of the Interstate Highway System for an equivalent amount of federal funds to be spent on the modernization of public transport. Judith Bender describes the coup de grâce for Westway as the refusal of the federal government in 1985 to allow an extension in the deadline for a trade-in of the $1.7 billion in federal funds earmarked for the project to invest in alternative highway and public transport improvements. Further obstacles emerged with the failure of the Army Corps of Engineers to win a twelve-year legal battle to begin construction work on account of potential damage to the decayed wharves, which serve as winter shelter for the striped bass, a commercially valuable migratory fish. The ecological arguments were in turn bolstered by growing constraints on public expenditure during the 1980s pushed by a more fiscally conservative Congress, and a deepening rivalry between the New Jersey and New York delegations which began to undermine the political coalition behind its construction.[84]

In 1985 the Westway proposal was finally abandoned, the victim of a politically incongruous coalition of environmentalists, mass transit lobbyists, and fiscal conservatives. This marked an effective end to the era of large-scale urban reconstruction in New York. "When Westway was declared legally dead," wrote Sam Roberts in the *New York Times,* "many of the eulogies also mourned the loss of New York's will to conceive and build great public works."[85] The roots of the Westway project can be traced to the regional plan of 1929, which had been conceived in an era when public opinion was subservient to technical expertise. With the reelection of Mayor Lindsay in 1969 the principle that planning decisions must involve community participation effectively prevented the construction of any more major highways in the city. Supporters of the Westway project such as Governor Mario Cuomo reflected on the changing context for public policy and the emergence of a new era of property rights, environmental rights, and expanded taxpayer rights in the 1980s.[86] Yet this change was due as much to the fiscal erosion of state power as to any belated recognition of public concern: in the 1970s the project had been defended as a spur to the city's economic regeneration; by the 1980s the scale of federal involvement it would have required was not compatible with the new neoliberal orthodoxy of limited public expenditure. The dissolution of the New Deal consensus on public works held implications for urban governance that went beyond the emerging public disquiet at grassroots levels. This is not to argue that federal policy had ever really been independent of the power of capital over this period, but that the relative autonomy of government institutions was now starkly reduced in comparison with the early 1970s when the Westway project had first been proposed.[87] With the refinancing of New York City in the late 1970s and the city's reentry into the municipal bond market, the scope for urban policy making had become far more restricted than in the past. The real estate boom of the 1980s seemed only to emphasize that municipal government had become a mere facilitator of broader processes over which it no longer had any effective control. And by the second term of the Reagan administration there were cuts across a range of federal budgets mirroring the earlier scaling down of urban policy under Nixon in the early 1970s.

Westway failed because the proliferating delays and legal challenges exposed its anomalous nature as a relic from a previous era. Most tellingly of all, the Westway project was fiercely resisted despite its sophisticated landscaping and design components. It is ironic that Westway, the most bitterly fought of the major highway schemes, sought to address the earlier concerns over the Cross Bronx Expressway, the Cross Brooklyn Expressway, and the Lower Manhattan Expressway. The road was to be sunk beneath a 93-acre waterfront park and provide 133 acres of new land for development, yet these design concessions made little impact on public debate. Robert Venturi and Denise Scott Brown's design for the Westway park was a postmodern theme park replete with medieval castles, miniature Empire State Buildings, and giant apples.[88] But their quirky design found little public favor. By the 1980s landscape architecture had drifted into disrepute as little more than a service industry for real estate, and the very idea of a landscaped parkway had become an anachronism.

3.4 FRACTURED CITIES

In the early decades of the twentieth century the urban landscape of New York City was radically altered to accommodate demands for greater freedom, leisure, and mobility. The development of the private automobile set in train an irrevocable dynamic of ever greater social and spatial separation that challenged nineteenth-century conceptions of urban planning. The growth of the suburbs was explicitly fostered by an appeal to a pastoral ideal that served to undermine the rationale for urban nature sustained by the legacy of nineteenth-century urbanism. With mass migration to the suburbs and easy access to the countryside provided by cars, the arguments for the provision of nature in cities lost their rhetorical power.

The extent of urban reconstruction in the New Deal era can be explained in part by the generous federal funding arrangements, which ensured that cities could embark on major projects without running into fiscal difficulties. With the multibillion-dollar Federal Highway Program and the vast suburban housing initiatives of the Federal Housing Administration, a new car-dominated urban landscape was created. The construction of highways accelerated the suburbanization of urban space and contributed to postwar urban decay. Leading figures in urban planning such as Lewis Mumford decried the postwar expansion of roads and warned of their impact on both cities and countryside alike. He derided "the bacteria of finance" that had wrought such destruction to the American landscape, but he could not articulate any clear alternative to capitalist urbanization.[89] Mumford was writing from another age, attempting to adapt the principles of the city beautiful movement and the Progressive era to a set of radically different circumstances; he misjudged both the growing American love of cars and the ever greater desire for private space under the consumerist trajectory of twentieth-century modernity. In contrast, Robert Moses clearly understood the growing centrality of the car to postwar America. In 1947, for example, he observed:

> Out of all the welter of controversy about postwar public works one fact is emerging. The average citizen is enthusiastic about highway improvement. . . . Every American wants a durable, cheap car, and he looks to the automobile industry to provide it. He has no desire to wear that car out quickly on a broken and obsolete road system. What is more, he is willing to pay the bill.[90]

The parkways of the early twentieth century worked to create a new authenticity in the American landscape through their combination of technological innovation and a search for a regional organicism in urban form. Although there are strong commonalities between the interwar experience of Europe and America, there are also significant differences in the perceived role of nature in urban form. Unlike Italian futurism, for example, the American modernist ideal was founded on a close synthesis between the transformation of nature and the iconography of machine technology.[91] American modernists such as Frank Lloyd Wright welcomed the dispersal of urban form enabled by technological change as a means to foster a closer interaction between society and nature. Yet this regional vision conflicted with the social and economic tensions emerging in existing cities: a nineteenth-century preoccupation with the garden city synthesis occluded any deeper engagement with the dynamics of twentieth-century urban form. In subsequent decades this interwar ecological modernism developed by Wright, Eckbo, Jensen, and others was gradually lost amid the corporate modernism of postwar reconstruction. "Now the modern revolt seems to have run its course," remarked Eckbo in 1978, "leaving a collection of promising precedents and incomplete theories."[92] The earlier conceptions of technology and landscape were supplanted by a profit-driven technocratic vision in which there was little space for a regional planning ethos.

The complexity of the relation between landscape iconography and the articulation of wider political ideals is rooted in a shifting and often contradictory relationship between aesthetics and the public realm.[93] Ultimately of course this uncertain connection between the modernist aspirations of a technical elite and the wider arena of popular culture, fostered by the momentum of postwar social and economic change, would problematize the very basis for any lasting synthesis between technological modernism and regional planning. And the collectivist ideals of modernity would conflict with the ever greater demands for private space epitomized by the emerging "autopia."

The legacy of technological modernism and its ultimate rejection illustrates how the modern movement failed to develop a workable set of relations between itself and nature.[93] Although the earlier achievements of Moses made much of the "urban pastoral" as a new kind of conjunction between technology and nature, the later projects, particularly after World War II, can be conceived as the most intense separation between nature and human artifice in the history of New York. Under Moses, the original concept of the urban parkway had become "deformed" and New York, "at first a champion of the parkway, later became its nemesis."[94] The later generation of multilane urban highways, such as the Gowanus Parkway in Brooklyn, were not landscaped roads in the earlier sense but ugly expressways, in this case suspended above Sunset Park by a line of steel columns (figure 3.7).[95] In one sense, the legacy of Moses can be read as symptomatic of the perceived hostility toward nature in modernist aesthetics, a divide

3.7 Gowanus Parkway under construction, Brooklyn, 1941.
Source: Photograph by Rodney McCay Morgan. MTA Bridges and Tunnels Special Archive.

between urban design and nature stemming from the demise of nineteenth-century planning ideals. This separation from nature under technological modernism is best conceived as an emerging tension between the public nature of nineteenth-century urbanism and the privatized nature emerging from the car-oriented Fordist era of mass consumption and new middle-class aspirations. What we find in the case of New York is a peculiarly intense concentration of the forces and contradictions within technological modernism and their physical expression through the ideals of comprehensive urban planning.

The development of landscaped roads involved a refashioning of aesthetic sensibilities toward both nature and urban space. The simple association of road construction with the destruction of nature ignores the subtle interplay between the different aesthetic and ideological responses to nature that have developed in tandem with the evolution of urban parkways and highways. The interwar American parkways sought to transform technology into architecture, science into art, and nature into culture; they represented a radical fusion of politics, aesthetics, and urban form.[96] Yet Moses and his public works increasingly embodied the contradictions of modernism and modernist aesthetics. By the 1950s cultural modernism drew little energy or inspiration from nature, in contrast to the earlier nature-based visions of Le Corbusier, Garrett Eckbo, and Frank Lloyd Wright. The eventual dissipation of the Moses era marks an effective end to comprehensive physical planning and the waning of state-directed attempts to rationalize the spatial order of American cities. With little sense of irony Moses compared his own achievements with Baron Haussmann's transformation of nineteenth-century Paris. The story of Haussmann is "so modern," wrote Moses, "and its implications and lessons for us so obvious that even those who do not realize that there were planners before we had planning commissions, should pause to examine this historic figure in the modernization of cities."[97] Yet the decline of Moses's influence marks the eventual exhaustion of this phase of urban history and the demise of the master planner as a credible figure in the transformation of the urban landscape.

The populist vision of Moses was at best only tangential to the earlier ideals of the American modernist movement. The failure of technological modernism does not necessarily lead us toward the antimodern Jeffersonian vision of Jane Jacobs or validate a generalized indictment of the modernist movement as a whole. As Marshall Berman suggests, the 1960s saw the emergence of a new, more politicized response to modernity, an alternative kind of modernity to the "expressway world" of concrete and steel.[98] As we shall see in the next chapter, the political and economic turmoil of the 1960s fostered a different kind of engagement between utopian ideals and the urban environment. The innate contradictions behind technological modernism became exposed to unprecedented scrutiny by marginalized urban voices who sought to challenge the use and meaning of urban space.

4

Between Borinquen and the *Barrio*

Our cry
is a very simple
and logical one.
Puerto Ricans came
to this country
hoping to get
a decent job
and to provide
for their families;
but it didn't take
long to find out
that the American
dream that was
publicized so nicely
on our island
turned out to be
the amerikkkan
nightmare

—*David Perez*[1]

Juan
Miguel
Milagros
Olga
Manuel
From the nervous breakdown
streets
Where the mice live like
millionaires
And the people do not live
at all
Are dead and were never alive

—*Pedro Pietri*[2]

In April 1996 a documentary film on a little-known Puerto Rican movement of the 1960s called the Young Lords was premiered at the Lincoln Center for the Performing Arts in midtown Manhattan.[3] After the screening, the director, Iris Morales, thanked the audience and invited the real figures from the documentary to step forward. One by one, former members of the Young Lords made their way down the cinema aisles and joined her to receive a prolonged standing ovation. Many members of the audience were young activists drawn from contemporary Latino organizations in the city who were curious to learn more about their past. The whole event was charged with a powerful sense of historical continuity between the civil rights struggles of the 1960s and a new wave of commitment to tackle the interrelated problems of poverty, racism, and pollution in the 1990s.

Recent years have seen an upsurge of both scholarly and popular interest in the United States environmental justice movement. This radicalization of environmental thinking presents a distinct political agenda quite different from established concerns with the protection of "wild nature" and narrowly defined conceptions of environmental quality. The environmental justice movement has succeeded in bringing issues of race, class, and gender to the center stage of the American environmental debate, yet we know little of the historical roots of this interaction between environmental politics and social justice. This chapter examines a specific element in New York's radical environmental tradition that emerged in the 1960s from the politicization of the city's Puerto Rican community in the face of systematic political, economic, and cultural marginalization. The origins of urban environmental struggles at this time reveal a radical fusion of grassroots demands for greater community control over urban space with a powerful emphasis on social justice garnered from the civil rights movement. We trace the emergence of the Young Lords, who successfully mobilized the Puerto Rican community through a series of direct actions devoted to improving living conditions in the ghetto, the creation of new community spaces, and the assertion of cultural identity. We examine internal tensions within the organization as it sought to extend its role beyond community-based concerns such as sanitation and health care, toward more abstract political goals focused on demands for Puerto Rican independence.

In this chapter we explore the conception of "nature" as a material and symbolic dimension to human experience that extends from the health of the human body to the social production of urban space. There is not one "nature" or "environment" to be controlled by technical elites in the service of an imaginary public but many "natures" coexisting amid the social and cultural diversity of the city. Our perspective is now very much from the street rather than from the office of the planner or the control room of the engineer. We can begin to explore some of the contradictory and contested dimensions to the transformation of urban space as part of the extraordinary ferment of social and political change that swept through American cities at this time.

The 1960s are perhaps the least understood and most contentious juncture in twentieth-century US history. This decade saw the most sustained challenge to the legitimacy of US society and its governing institutions since the widespread labor unrest of the Great Depression. The tensions underlying the New Deal and the unequal distribution of postwar prosperity exploded into violent protests across the ethnic ghettos of over two hundred US cities. Urban America became the focal point for a critical mass of discontent fanned by racism and police harassment, undemocratic and unresponsive local government, declining economic opportunities, cultural invisibility, poor housing, and the disruptive effects of urban renewal. Out of this era emerged a complex array of community-based struggles demanding an end to the injustices and exclusions of the past.[4] The civil rights movement and the paradoxical legitimation of social change through the federal antipoverty programs of the Kennedy and Johnson era had suddenly and dramatically raised the political expectations of many millions of Americans who had hitherto been systematically excluded from both material prosperity and social empowerment.[5]

One of the most innovative groups to emerge from this wave of political activism was the radical Latino organization called the Young Lords, led principally by second-generation Puerto Ricans, many of whose parents had been lured to the United States during the labor recruitment campaigns of the 1940s and 1950s. The Young Lords first emerged in Chicago during the mid-1960s and subsequently campaigned in New York City, Philadelphia, Newark, Hoboken, and a number of other American cities. The Young Lords, along with other Latino groups like Los Siete de la Raza, the Brown Berets, and Por los Niños, formed an integral part of this changing political scene.[6] The Latino contribution to urban environmental history has long been neglected in relation to the dominant narratives of city beautification, early public health reform, and the origins of urban planning. Yet those earlier interventions in urban space, as we shall see, were highly uneven in their impact. For millions of the urban poor the very idea of the urban environment was a bitter evocation of blight and neglect. Overcrowded and dilapidated tenements housed whole communities who had little chance of sharing in America's postwar prosperity.

By 1970 it is estimated that over 10 percent of New York City's population was of Puerto Rican origin. Yet despite their demographic strength they remained systematically underrepresented in virtually every sphere of decision making in the city.[7] This chapter charts a dramatic attempt to gain political visibility in the late 1960s and early 1970s through the building of a Puerto Rican social movement devoted to the transformation of their city. Though the Young Lords movement failed to achieve a permanent presence within the city's social and political structure, they prefigured a series of shifts in the social complexion, cultural identity, and political agenda of urban America. Their radical political agenda succeeded in exposing a series of injustices and contradictions in the urban and environmental legacy inherited from the New Deal era and its earlier precursors in the moral and technical discourses of nineteenth-century urbanism.

4.1 Landscapes of Despair

> Nothing but American thrift and industry will develop the agricultural and mineral resources of these islands.
>
> —*Charles H. Rector*[8]

The earliest interactions between New York City and the Spanish colony of Puerto Rico were based on trade for commodities such as sugar, rum, tobacco, and molasses. From the mid-nineteenth century onward, earlier trade-based patterns of Puerto Rican migration to New York were supplanted by a new wave of immigrants who were predominantly political exiles opposed to the continuing Spanish occupation of the island. New York quickly emerged as an important center for supporters of El Grito de Lares, a declaration of independence that lay behind the unsuccessful Puerto Rican rebellion against Spanish rule in 1868. By the 1890s radical Puerto Rican groups in the city had forged close links with the movement for Cuban independence and a variety of other organizations committed to Antillean liberation.[9] Following the end of the Spanish-American War in 1898, the island of Puerto Rico became a dependent territory of the United States and most of these radical groups disbanded. Within two years United States military authority had been replaced by Washington-appointed civilian governors and a rudimentary legislature was established for the island. Early descriptions of the island could scarcely contain their enthusiasm for America's new colonial possession. One commercial handbook described Puerto Rico as "one of the most lovely of all those regions of loveliness which are washed by the Caribbean Sea; even in that archipelago it is distinguished by the luxuriance of its vegetation and the soft variety of its scenery."[10] Another early guide to the island remarked how the entire landscape was "wonderfully crowned with low-hanging, vaporous clouds, which roll forever into new, fantastic, nebulous forms."[11] Beyond the aesthetic satisfaction, however, lay a determination to fully exploit the island and its people in order to realize its commercial potential: "The industrial possibilities of the island of Porto Rico, considering the fertility of its soil, the mildness of its climate, the abundance of rain, its insular position, and its teeming population can scarcely be viewed with anticipations too optimistic."[12]

At first, the Americans were welcomed as liberators against Spanish rule, but as the geopolitical, cultural, and economic significance of the US presence became clearer, deep political divisions emerged on the island that were to dominate Puerto Rican politics throughout the twentieth century. It was reported to the American House of Representatives as early as April 1900 that "syndicates to buy up practically all of the rich sugar, tobacco and coffee lands of the island were already being organized, and that representatives of the great railroad, telegraph and other corporations had been besieging Congress for months for the purpose of obtaining lucrative concessions."[13] Under American control the island's economy was transformed from a diversified, largely subsistence economy into a monocultural sugar-based economy predominantly under the control of absentee US owners, who quickly gained control of half the island's productive soil.[14] Even Rexford Tugwell, the Roosevelt-appointed governor of Puerto Rico, had titled his account of the island *The Stricken Land*. "Wages were miserable," wrote Tugwell, "living costs high, housing as bad, surely, as any in the world, people half starved on a diet inherited by tradition from the days of slavery, and sick with all the diseases to which malnourished and ill-housed people are subject in the tropics."[15] By the early decades of the twentieth century, most Puerto Ricans were arriving in New York for economic rather than political reasons: between 1909 and 1916, for example, over 7,000 Puerto Ricans arrived in the United States to escape agricultural ruin in the sugar and coffee industries. Yet New York City was to prove no haven, as the new arrivals found themselves the targets of extensive racial violence. In response to the marginality and vulnerability of their community, the Puerto Rican Brotherhood of America (La Hermandad Puertorriqueña) was founded in 1923 as a political lobby, and the Puerto Rican and Hispanic League for Civic Defense (La Liga Puertorriqueña e Hispana) was founded in 1926 for protection against violent attacks. Whereas La Hermandad Puertorriqueña was the first predominantly working-class organization to directly address the living conditions of Puerto Ricans in New York City, La Liga took an even more radical step of seeking to unite all Spanish-speaking migrants under one political organization. During the Depression years radical political activism also developed through organizations such as Club Pomarosas, Club Eugenio María de Hostos, and the American Communist Party.[16]

The depression of the 1930s brought a near collapse in the Puerto Rican economy marked by escalating poverty, malnutrition, and despair. Political tensions between the island's elite (dominated by sugar and coffee plantation owners) and the rest of the population intensified, with widespread violence and unrest. In the 1940s and 1950s emigration was actively encouraged by the ill-fated modernization program named Operation Bootstrap instituted by Luis Muñoz Marín and his governing Popular Democratic Party. The island was widely portrayed, by politicians and academics alike, as intrinsically backward and incapable of any form of self-determination for a combination of cultural, social, and environmental reasons. Yet the attempt to achieve rapid industrialization under Operation Bootstrap did not solve the island's problems but led to even higher levels of agricultural unemployment and greatly strengthened political, cultural, and economic ties to the United States.[17] As a consequence of these developments, the number of Puerto Ricans in New York grew rapidly between 1940 and 1960 from around 60,000 to over 600,000, with the new arrivals concentrated in three parts of the city with well-established Puerto Rican communities: the South Bronx, the Lower East Side, and particularly the Latino ghetto of East Harlem, which became known as *el barrio* (figure 4.1).[18]

In the 1950s we can find a variety of upbeat assessments of the prospects for Puerto Rican migrants. In 1958, for example, Elena Padilla wrote of the Puerto Rican "good life" in America with better wages, better educational opportunities, better medical care, and many other advantages over life back on the island.[19] It was widely anticipated that New York's new arrivals would be integrated into the fabric of the city just as earlier waves of immigration had been since the nineteenth century.[20] In the 1960s, however, this earlier sense of optimism was rapidly dissipated by structural changes in the US economy that aggravated the marginalization of the Puerto Rican community. The sudden and intense industrial decline of the New York-New Jersey regional economy from the mid-1960s onward had a devastating impact on Puerto Rican workers, who were disproportionately concentrated in low-wage unskilled occupations such as the garment industry, cigar making, and transportation.[21] The existing Puerto Rican organizations that had emerged to defend their community's interests, such as the Puerto Rican Association for Community Affairs (1953), the Puerto Rican Forum (1958), and Aspira (1961), proved powerless to prevent a deterioration in Puerto Rican living conditions.[22] In particular, the Puerto Rican Forum and other organizations had played a key role in the early 1960s in the design of a comprehensive Community Development Project for the *barrio* under the auspices of the federal War on Poverty programs initiated by President Johnson. In the event, however,

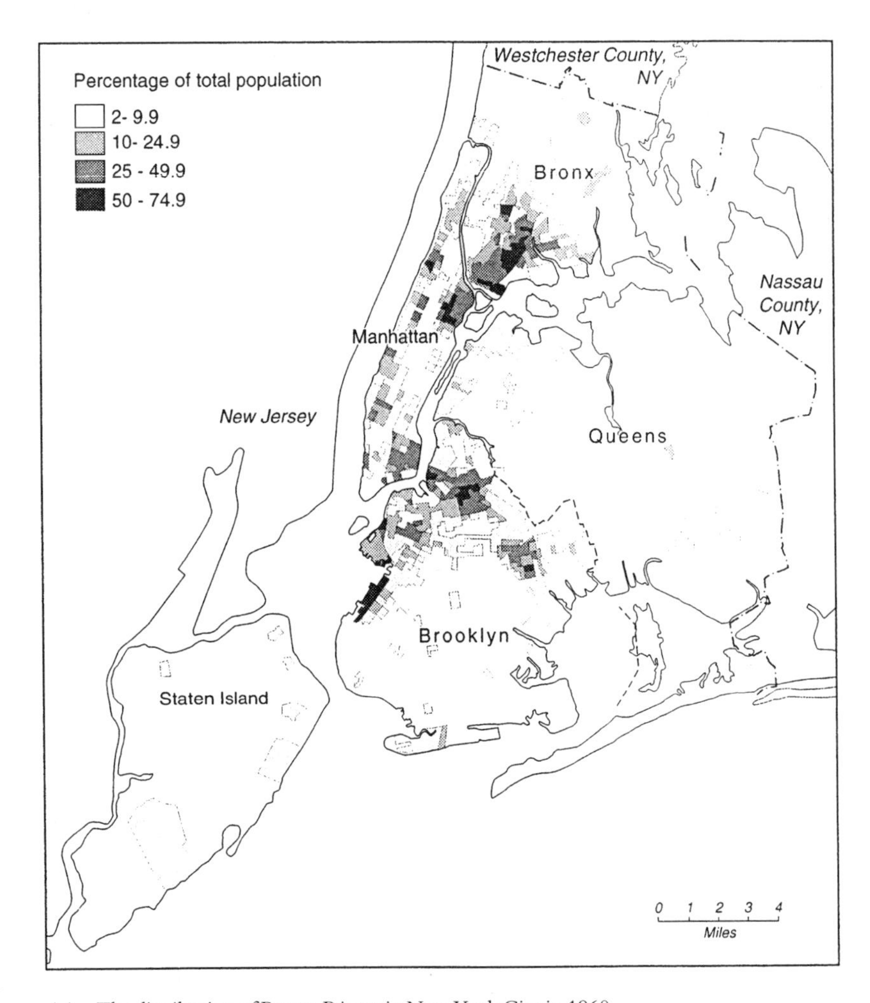

4.1 The distribution of Puerto Ricans in New York City in 1960.
Source: Data derived from Nathan Kantrowitz, American Geographical Society: Studies in Urban Geography No. 1 (1969).

the ambitious plans were never realized amid recriminations and political infighting, thereby opening the way for more radical alternatives to federal antipoverty programs. The failure of collaborative efforts between government agencies and established community leaders, combined with a renewed political emphasis on the US colonial legacy in the Caribbean, fostered the emergence of a new range of organizations in the 1960s such as the Movimiento Pro-Independencia (later Partido Socialista Puertorriqueño), El Comité, Unión Estudiantil Boricua (the Puerto Rican Student Union), and the Young Lords.

The loss of manufacturing jobs and other blue collar employment to the urban periphery, to the southern states, and to the newly industrializing nations was compounded by the impact of urban renewal and highway construction which fragmented and displaced working-class communities throughout the city. By 1968, with the election of the first Nixon administration, the earlier promise of Lyndon Johnson's Great Society and the prospects for wide-ranging social reform were rapidly displaced by escalating economic and political pressures to realign the balance of power between capital and labor in American society.[23] The images of American life promoted in the postwar labor recruitment campaigns in Puerto Rico were increasingly at variance with the new social and economic realities. By the late 1960s Puerto Rican unemployment in the *barrio* of East Harlem neared 40 percent, and a third of Puerto Rican families were below the poverty line.[24] The new arrivals found that they were restricted to menial, poorly paid work (if they found work at all) and to overcrowded, dilapidated housing and inadequate access to basic services such as education and health care. Widespread racial, cultural, and language discrimination completed the bleak outlook.

In the mid-1960s these simmering grievances led to violent protests by Puerto Rican communities across the United States. In June 1966 there was a major disturbance over four days involving 70,000 Puerto Ricans in Chicago after a police shooting incident. In August 1966 there were widespread clashes between police and Puerto Rican youths in Perth Amboy, New Jersey. And in July 1967, in the wake of the devastating Detroit riots, mass disturbances spread through East Harlem in Manhattan. As night fell on 24 July, thousands of Puerto Rican youths swept through the *barrio* in fierce antipolice demonstrations that left two people dead and many injured, in the city's worst civil disorder since the Harlem and Bedford-Stuyvesant race riots of 1964. At its peak the rioting extended all along Third Avenue from 119th to 103rd streets and between Park Avenue and Second Avenue, before 1,000 police reinforcements managed to contain it.[25] A widely held view in the early 1960s, reinforced by both media stereotypes and academic treatises, was that the Puerto Rican community was docile and politically inactive: the riots of the 1960s swept this misconception away overnight.[26]

4.2 Space, Identity, and Power

> The Young Lords is, was, the most important political event to happen for my generation of Puerto Ricans in recent history. The Young Lords were an initiation for a whole generation of people into their culture and their history.
>
> —*Eduardo Figueroa*[27]

The origins of the Young Lords Organization can be traced to a series of developments in Chicago during the 1960s. The Young Lords had first emerged as a street gang in the Lincoln Park area of the city in 1959. In 1968, under the leadership of José Jiménez, the Young Lords were transformed into a political organization dedicated to fighting social injustice.[28] In June 1969 the Chicago-based Black Panthers formed an alliance with a number of other radical groups in the city—the Young Lords among them—called the Rainbow Coalition, a "vanguard of the dispossessed."[29] This was to be one of the first multiracial political forums in the United States and was based on a set of shared concerns with social justice, community empowerment, and opposition to the Vietnam War. From the outset, the Young Lords emulated the Black Panthers in their emphasis on recruiting "street people" rather than factory workers. They also followed the Panthers in adopting a quasi-military structure with a central committee, ministers, and the wearing of easily recognizable purple berets. Other important influences on the

Lords at this time were the radical labor struggles led by the farm workers' union under the leadership of Cesar Chavez, the League of Revolutionary Black Workers at the Detroit Chrysler plant, and the international liberation movements in Asia, Africa, and Latin America.[30] The emergence of the Young Lords, along with a number of other radical organizations such as the Puerto Rican Socialist Party, El Comité, and the Puerto Rican Students Union, heralded a far more combative stance toward the poverty and racism facing their community than the political and social organizations established by first-generation migrants. Like the Panthers, the Chicago-based Lords had their own newspaper and were engaged in several community-based campaigns, such as opposition to urban renewal, and in programs to distribute free breakfast and free clothing; they had also occupied a church to press their demands.

Through their adept handling of the media they soon caught the attention of Puerto Rican activists in New York City, where a chapter of the movement was founded in July 1969.[31] As the New York activist Pablo "Yorúba" Guzmán recalls:

> What happened was, in 1969 in the June 7 issue of the Black Panther newspaper there was an article about the Young Lords Organization in Chicago with Cha Cha Jimenez as their chairman. Cha Cha was talking about revolution and socialism and the liberation of Puerto Rico and the right to self-determination and all this stuff that I ain't *never* heard a spic say. I mean, I hadn't never heard no Puerto Rican talk like this—just Black people were talking this way, you know. And I said, "Damn! Check this out." That's what really got us started. That's all it was, man.[32]

The New York chapter started recruiting members by forging links between various street gangs from East Harlem and the Lower East Side and student activists from the recently created Sociedad de Albizu Campos. From the outset, the movement clearly differed from existing neighborhood-based groups in the city devoted to issues surrounding planning and urban renewal. The activists were drawn from diverse backgrounds: students, ex-gang members, ex-convicts, addicts and other street people. The Lords were indisputably a youth group, with most of their members and supporters in their late teens or early twenties. They were also overwhelmingly second-generation Puerto Ricans, the *Nuyoricans,* whose life experience and political consciousness were shaped in the urban *barrios* of America rather than in the poverty-stricken sugar and coffee plantations of the island. The organization was rooted in a rediscovery of Puerto Rican identity and reflected the full diversity of the island's ethnicity with its mix of Iberian, African, and native Taino Indian ancestry. Yet the Lords also welcomed many supporters from outside the Puerto Rican community: as many as a quarter of their supporters were African-American, and there were significant numbers of Dominicans and Panamanians as well.[33] The racially and ethnically diverse Puerto Rican community had historically occupied an ambiguous space within the dominant black-versus-white racial antinomy of the city. Differences in class and race among Puerto Ricans had combined to produce a complex hierarchy of identifications within American society, driven by an assimilationist desire to be fully integrated into white society. Though radical groups in the 1920s such as La Hermandad and La Liga had sought to challenge US society and build radical political alliances, they had lacked the broad-based momentum for political change that flowed in the wake of the civil rights movement. The Lords effectively inverted existing racial identification by insisting on the Puerto Rican difference from white America and drew attention to the powerful commonality of interests with the burgeoning Black Power movement. In this sense the Lords gained political impetus from the long-standing political and cultural interchange between Latinos and African-Americans in New York and fiercely rejected the assimilationist aspirations of middle-class Puerto Ricans.[34]

One of the first activities of the newly formed Young Lords was to find out what issues were of greatest concern to the Puerto Rican community. To the evident surprise of some Lords activists, the most immediate preoccupation turned out to be the filthy state of the streets in the *barrio.* Piles of garbage were being routinely ignored by the city's sanitation department, in stark contrast to the pristine sidewalks of affluent districts in downtown Manhattan. For the residents of the *barrio,* uncollected garbage had become a poignant symbol of the indignity of poverty, political invisibility, and municipal neglect. A typical *barrio* landscape is described in the Lords newsletter *Palante:* "East Harlem is known as El Barrio—New York's worst Puerto Rican slum. . . . There is glass sprinkled everywhere, vacant lots filled with rubble, burnt out buildings on nearly every block, and people packed together in the polluted summer heat. . . . There is also the smell of garbage, coming in an incredible variety of flavors and strengths."[35] These concerns led to the Lords' first direct action, the so-called "garbage offensive," which was launched on 27 July 1969. As Guzmán recalls: "The best thing to hook into was garbage, 'cause garbage is visible and everybody sees it. . . . So we started out with this thing, 'Well, we're gonna clean up the street.' This brought the college people and the street people together, 'cause when street people saw college people pushing brooms and getting dirty, that blew their minds."[36] Within a few weeks there were so many people involved that a group of activists demanded extra brooms from the local sanitation department in order to extend the mass sweeping of the streets, but the dismissive attitude of the city's sanitation service toward the Lords' demands for clean streets led to a rapid escalation in the garbage activism. In order to draw attention to the state of their neighborhood, the Lords activists and supporters then began repeatedly placing uncollected garbage in the middle of the streets overnight in order to force the city authorities to take action. By 17 August the action spread as uncollected garbage and burning cars were used to barricade the main avenues in the *barrio:* Madison, Lexington, and Third avenues were blocked at 110th, 111th, 115th, 118th, and 120th streets, and across a six-block area angry residents staged what the *New York Times* described as a "garbage-throwing melee."[37] Meanwhile, a crowd of 1,000 people marched on the police station at 126th Street to draw attention to widespread police harassment. The day after the rally, one of the Lords' most high-profile activists, Felipe Luciano, who had been a founding member of a radical black theater group called the Last Poets, reflected on the symbolic importance of the garbage offensive:

> They've treated us like dogs for too long. When our people came here in the 1940's, they told us New York was a land of milk and

> honey. And what happened? Our men can't find work. . . . Our women are forced to become prostitutes. Our young people get hooked on drugs. And they won't even give us brooms to sweep up the rubbish on our streets.[38]

The garbage offensive provided the Lords with a highprofile community presence and attracted many new supporters. The state of the *barrio* had for years served as a symbolic reminder of the community's marginalization, a daily indignity that led either to soul-destroying passivity or to real anger. In a sense, the Lords and their young supporters were carrying out an action in defense of their parents' generation. By publicly sweeping the streets, they had extended the idea of home into the wider space of the city. With the construction of trash can barricades, the demand for clean streets had become a revolutionary act. The politicization of garbage also challenged the racialization of hygiene inherited from nineteenth-century conceptions of medical science which reinforced popular prejudices concerning dirty living and bad neighborhoods. Hundreds of thousands of Puerto Rican families were forced to occupy the dilapidated pre-1901 "old law tenements," which had decades earlier been endured by Irish, Italian, and Jewish settlers (figure 4.2). "Since most Puerto Ricans in New York City live in slum conditions," wrote Dan Wakefield in his classic survey of the late 1950s *barrio*, "it is part of the popular myth of their migration that they created the slums. The truth is, rather, they inherited the slums. . . . Those dilapidated buildings were the ghosts of the country's first great waves of immigration."[39] For a while, the city's sanitation department responded to community demands, and the *barrio* was as clean as any middle-class neighborhood. Mayor John Lindsay attempted to resolve the conflict through the use of existing contacts based in government antipoverty programs, but mayoral representatives were met with widespread hostility. There is little doubt that the most senior Puerto Rican politicians in the city at the time, such as Roberto Garcia, Ramón Vélez, Gilberto Gerena Valentín, and Hernán Badillo, were held in contempt by many of the Lords activists.[40]

4.2 Poverty and the social environment: Cauldwell Avenue, Bronx, New York, November 1970.
Source: Photograph taken by Michael Abramson. Reproduced with permission from Michael Abramson, *Palante: Young Lords Party* (New York: McGraw-Hill, 1971).

In October 1969, in a further emulation of the Black Panthers, the Young Lords launched their Thirteen Point Program, which set out the organization's political objectives. The Lords began to shift their attention to a wider array of campaigns involving the provision of day care for single mothers, the cooking of free meals for children, and the monitoring of tuberculosis, yet they lacked access to any building large enough to enable them to carry out their work effectively. A delegation from the Lords requested the use of the First Spanish Methodist Church on the corner of 111th Street and Lexington Avenue as a base from which to run their breakfast programs. Although the church stood empty for most of the week, both the pastor (a fiercely anticommunist Cuban exile) and the church board refused to allow the Lords to use the building, referring to the activists as "Satanás—the devil."[41] On Sunday 7 December 1969 some activists attempted to address the congregation but were badly beaten by a large number of police. Former activist Luis Garden-Acosta recalls this incident as his first encounter with the Young Lords:

> I heard about these young people who were trying to get a church which was not used or open much during the week to use that space for a breakfast program and clothing drives. . . . The church had allowed police to come in during a service and blood was spilt. I mean that's really grave—repugnant—allowing police to come into the church. . . . So when I heard about that—next Sunday I was there.[42]

On 28 December there was a second attempt to use the church after the Sunday services. In a carefully planned action, over a hundred Lords activists and their sympathizers succeeded in occupying the building (figure 4.3). A wooden sign with red lettering proclaimed that the building was now *La Iglesia de la Gente*—People's Church. The church board immediately sought an injunction in the State Supreme Court demanding that the Lords and their supporters vacate the church, but the occupiers refused to comply. As the preliminary injunction was read out on the church steps to a crowd of supporters, activists handed out leaflets stating that "the first responsibility of the church is to the people." Splits quickly

4.3 Marching to the People's Church through the streets of East Harlem, December 1969.
Source: Photograph taken by Michael Abramson. Reproduced with permission from Michael Abramson, *Palante: Young Lords Party* (New York: McGraw-Hill, 1971).

emerged within the citywide church hierarchies; a group of white seminarians accompanied by college students (drawn principally from the Students for a Democratic Society based at Columbia University) began a sit-in at the Methodist Board of Missions on the Upper West Side to protest against the church's use of the courts against the Young Lords. The church occupation also created a public rift among city politicians: the leader of the East Harlem Task Force, Arnold Segarra, publicly backed the Young Lords by reading out a declaration of support by community leaders drawn from across the city and was promptly sacked by Mayor Lindsay.[43]

Over a period of eleven days some three thousand people visited the church to participate in the breakfast programs, the free clothing drives, a community school, a day care center, free health care, and a range of cultural activities including music and poetry. A series of educational events billed as "liberation classes" were also held, in which the history of Puerto Rico was combined with readings from the work of Pedro Albizu Campos, Eldridge Cleaver, Malcolm X, and Che Guevara.[44] Whereas the garbage offensive had attracted mainly local attention within the *barrio,* the church occupation drew not only citywide but also national and international media interest.[45] On 7 January 1970 the police regained control of the church and 105 people were arrested:

> At 6:30, as the snow fell lightly, hundreds of helmeted police officers took up positions around the peak-roofed church building and sealed off traffic in the surrounding streets. . . . Informed of their arrest, the occupiers filed out in groups of 20 into waiting police vans and buses. Some shouted "Power to the people," as they walked down the steps and others sang "Que Bonita Bandera" ("What a beautiful flag"), a Puerto Rican folk song. . . . After the arrests, the defendants were taken to the courtroom of Justice Saul S. Streit, administrative judge of the State Supreme Court, who had signed the arrest order at 10 a.m. As their names and addresses were called off, the Young Lords rose, many of them correcting the reader by giving the Spanish pronunciation of their names.[46]

Following the end of the church occupation, the Lords turned their attention to inadequate health care in the city. Poverty, poor housing, and lack of access to basic medical services had combined to produce high rates of tuberculosis, lead poisoning, and a variety of other "diseases of oppression" in their community (figure 4.4). An initial focus of their health campaign was the pervasive incidence of tuberculosis in the crowded tenements of the *barrio*. As early as 1938, for example, Lawrence Chenault had noted that tuberculosis was the most serious health threat facing the city's Puerto Rican community. High rents, dilapidated buildings, and overcrowding had contributed toward higher rates of the disease among New York's Puerto Rican community than in the slums of Puerto Rico. Furthermore, ill-judged public health campaigns in the 1930s led by federal and city authorities had succeeded in stigmatizing Puerto Ricans as carriers of disease in the absence of any significant advances in living conditions. In 1934, for example, Eleanor Roosevelt had declared to the New York Women's Trade Union League: "I assume that none of you will be hiring any of them in your homes, but however careful we may be in rearing our children, they can still come into contact with one of those sick people in the streets or in the schools." We can interpret the tuberculosis campaigns of the Lords as a continuation of the interwar efforts of, for example, the Puerto Rican and Hispanic League and its newspaper *El Nuevo Mundo* to demand improved public health facilities for their community and challenge the racialized representations of disease by official agencies.[47] The Lords activist Gloria Gonzales describes their efforts to extend the screening of tuberculosis:

> We started going door to door and doing the tine test, which is a very simple thing—the myth that only doctors could do it was done away with—and found again that a lot of our people were suffering from TB. . . . Finally we ripped off a TB truck. We knew the city had this truck that was used in *El barrio* only two hours a day. We wanted to use it around the clock to serve our people. They said no, we couldn't have it, that it didn't belong to them, that they had rented it. The thing was, our people couldn't wait, 'cause they were dying from

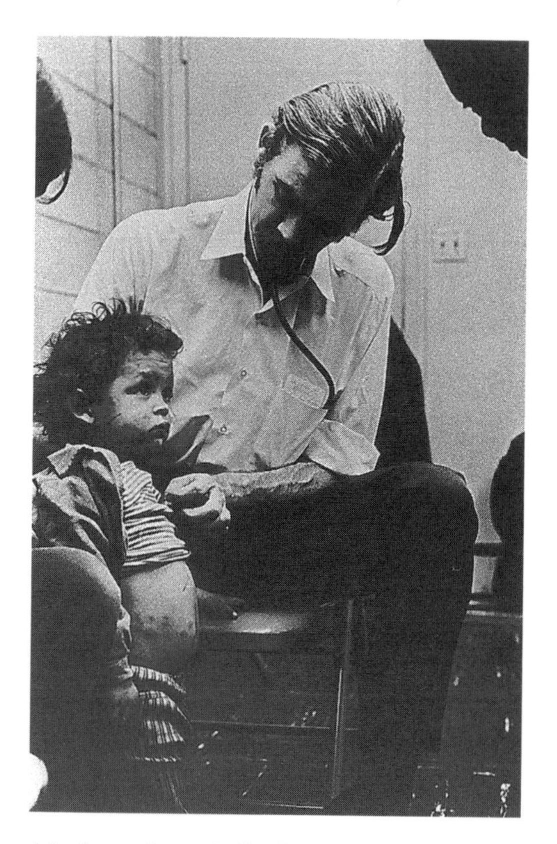

4.4 A case of severe bedbug bites: free health clinic at the First People's Church, December 1969.
Source: Photograph taken by Michael Abramson. Reproduced with permission from Michael Abramson, *Palante: Young Lords Party* (New York: McGraw-Hill, 1971).

> TB. We took the truck and we started testing our people, and in one day we tested 1,000.[48]

The politicization of health care not only focused attention on environmental inequalities within the city but also necessitated an examination of the relationships between gender, health, and social power. Although there was little input from women when the Young Lords first began campaigning, women quickly emerged as the main beneficiaries of the free day care facilities, the clothing drives, and the provision of free health care by doctors and nurses from radical medical organizations in the city. A range of urban scholarship has revealed how the role of women within urban social movements stems from consumption-oriented issues based in civil society, rather than those of the male-dominated workplace of the Fordist era.[49] By 1970, women held key leadership positions in Lords branches in the Bronx, East Harlem, the Lower East Side, Newark, and Philadelphia, and estimates suggest that women now made up at least 40 percent of the active membership (figure 4.5).[50] Some sense of the range of inspiration for radical Puerto Rican feminism at this time can be captured in the diverse array of women activists celebrated in the Lords newsletter *Palante,* including Mariana Bracetti (of El Grito de Lares), Lola Rodriguez de Tio (an imprisoned poet and political activist), Sojourner Truth (born a slave in 1797 in Ulster County, New York, who traveled around the North in the 1840s to speak against slavery and for women's rights), Angela Davis (of the Black Panthers), Blanca Canales (leader of the Jayuya uprising in 1950 in Puerto Rico), Lolita Lebron (the Puerto Rican activist jailed for the 1954 attack on the US House of Representatives), and the Vietnamese military leaders La Thi Thim and Kan Lich.[51]

The Young Lords activists drew attention to the fact that Puerto Rican women, as a result of inadequate prenatal and postnatal care, were experiencing significantly higher levels of infant mortality than the city as a whole.[52] The sense of indignation was also fueled by the death of a leading Lords activist, Carmen Rodriguez, after an abortion at the ill-equipped Lincoln Hospital. The Lords' third direct action in July 1970 was to be the occupation of this dilapidated public

4.5 Iris, Denise, Lulu, Nydia, May 1970.
Source: Photograph taken by Michael Abramson. Reproduced with permission from Michael Abramson, *Palante: Young Lords Party* (New York: McGraw-Hill, 1971).

hospital, which served 300,000 mainly Puerto Rican and African-American people in the South Bronx.

In the early hours of 15 July 1970 a group of 150 activists invaded the administrative buildings of Lincoln Hospital, occupying them for twelve hours. The hospital's newly appointed director, Dr. Antero Lecot, subsequently conceded that the dramatization of the hospital's plight had probably been helpful. As part of their health campaign the Lords set out a ten-point health program demanding self-determination in all health services and access to free health care. The occupation of the hospital also involved collaboration with a number of other newly formed organizations such as the citywide Health Revolutionary Unity Movement (founded by Lords activist Gloria Gonzales) and the Bronx-based Think Lincoln Committee of hospital workers and patients.[53]

The Young Lords were one of the first political groups in America to draw attention to the issue of sterilization administered without consultation to African-American and Puerto Rican women in public hospitals.[54] The issue of Puerto Rican fertility had wider implications: the poverty of the Puerto Rican community both in the United States and on the island had been repeatedly blamed on overpopulation. Virtually all studies of Puerto Ricans were preoccupied with this demographic context to the exclusion of any wider consideration of the colonial history of Puerto Rico. President Roosevelt, for example, had demanded in the early 1930s that "the frightening increase of the population had to be stopped."[55] And Nathan Glazer and Daniel Patrick Moynihan, in their influential 1963 study of New York migration *Beyond the Melting Pot,* had described how "misery among Puerto Ricans" is related to the high birth rate, in a crude form of demographic determinism.[56] Ominously, from the 1930s onward, Puerto Rico had become a kind of large-scale human experiment in mass sterilization and the testing of new birth control technologies. At the height of the sterilization program more than a third of the women of child-bearing age on the island had been operated on in one of the most heinous programs of neo-Malthusian "social engineering" of the modern era.[57] The radicalization of health care thus struck a powerful political chord that connected colonial medical practices and community marginalization.

The focus on women's health issues with the occupation of Lincoln Hospital had also emerged from a critical examination of gender inequalities within Puerto Rican society. Women activists drew attention to the way in which Puerto Rican wives were the target of extensive male violence as well as suffering the indignity of the traditional mistress or *corteja* system. The Lords sought to reform their community from within with the campaign slogan *La mujer puertorriqueña es doblemente oprimida* (The Puerto Rican woman is doubly oppressed). Newly emerging feminist ideas served to emphasize the simultaneous struggles against capitalism and *machismo* within their own community. In 1970 the call for "revolutionary machismo" was dropped from the Lords' Thirteen Point Program and replaced with an explicit commitment to equal rights for women. The growing involvement of women in the Young Lords represented a challenge to the existing gender roles of Puerto Rican society and the pressures of integration into the urban labor market for newly arrived migrant workers. Women were expected to fulfill traditional family roles yet were often the main breadwinners in Puerto Rican households through their employment in restaurants, laundries, the garment trades, and cigar factories. Though Puerto Rican women had played a crucial role in the New York economy since the 1920s, particularly within the garment industry, they suffered poor working conditions and widespread union discrimination. The economic restructuring of the 1960s and 1970s was to have a profound impact on the prospects for working-class women in the city and pushed countless families into poverty.[58] The Lords' interventions in debates over gender, work, and changing household structure provided a series of prescient insights into the dilemmas facing the Puerto Rican community in response to widening demands for social emancipation. Yet the Lords were faced with a growing paradox of heightened community mobilization at a time of accelerating urban decline, a tension that would prove irresolvable by the early 1970s. The advances made in terms of gender equality would prove fragile and ephemeral in the face of pressures to "professionalize" community leadership in the 1970s, and even experienced women activists from the Lords complained of being excluded from wider involvement in the Puerto Rican left.[59] The difficulties faced by the women activists in the Lords reflected the pervasive impact of autocratic styles of leadership within the civil rights movement and the degree to which the liberation movements of the time were highly uneven in their impact on gender inequalities.[60]

From the summer of 1969 until the spring of 1971 the Young Lords played a significant role in fostering a vibrant cultural and political sphere within the Puerto Rican community through high-profile direct actions, their newsletters and regular radio broadcasts, poetry readings, and other activities. They extended existing conceptions of the urban environment in multiple ways to encompass the control of community space, the creation of a healthy city, and the transformation of social relations. Their words and actions had revealed, for a brief moment, a different kind of urbanism springing from the injustice of the ghetto. In the early 1970s, however, their influence was to be rapidly dissipated by a combination of developments.

4.3 Disarray in the 1970s

> It's winter; winter in America
> and ain't nobody fighting because
> nobody knows what to save.
>
> —*Gil Scott-Heron*[61]

> The history of the 1960s and early 1970s has been expurged.
>
> —*Denise Oliver*[62]

In the first two years of Lords activity, the emphasis on Puerto Rican nationalism had been principally a means to foster a shared demand for greater political power and cultural self-determination. From 1971 onward, however, the overriding political aim of the Young Lords became Puerto Rican independence rather than redressing local grievances. The cultural nationalism of the *barrio* became more geopolitical in its focus and looked to the mythical island of Borinquen (Puerto

Rico in its imagined precolonial state) rather than the tangible transformation of the ghetto. This shift in emphasis recalled long-standing connections between Puerto Rican socialist ideas and demands for Puerto Rican independence: at its founding conference in 1938, for example, the Fourth International had called for the island's liberation from American control. With their increasingly internationalist stance, the Lords activists began to conceive of their own struggle as an outlier of the Third World liberation struggles. The *barrio* was simply an internal colony within a larger exploitative system of exclusion and dependency. "We find a Latin American proletariat in that huge New York City," wrote Dardo Cúneo in 1968. "It is economically dependent, culturally discriminated. It is in that Latin American proletariat of New York that, truthfully, Latin America begins."[63] The political outlook of the Young Lords was now increasingly linked to the geopolitical aim of Antillean liberation in the face of tightening political, economic, and cultural ties between Puerto Rico and the United States. The island of Puerto Rico was portrayed in *Palante* and other radical literature as ravaged by colonial rule: its agriculture in ruins, its mineral resources plundered, and the landscape blighted by foreign-owned industry. The precolonial island of Borinquen had become a kind of mythical homeland in the nationalist imagination, a lost paradise "all pregnant with sweetness," in contrast with the humiliating stench of the ghetto.[64]

In October 1970 there was a major independence demonstration in New York to commemorate the Puerto Rican rebellion of 1950 and a second (this time armed) occupation of the Methodist Church on 111th Street to protest against the death of a Lords activist, Julio Roldan, who was found hanged in police custody. At the 1971 Puerto Rican Day Parade in New York there was a riot between hundreds of Lords sympathizers and the police in which scores of people were injured, including nineteen police officers. The growing violence associated with Lords activity began to alienate public support and provoked an intensification of internal debates about their political agenda.[65] The emphasis on Puerto Rican independence highlighted deep ideological divisions within the Puerto Rican community, which were revealed by the city's annual Puerto Rican Day Parade. The organization of the parade (which remained under the control of the assimilationist establishment) exposed tensions between Puerto Rican ethnicity as simply cultural identity (linked to pluralist conceptions of urban power brokerage) and the concrete concerns of grassroots and pro-independence organizations committed to emphasizing the structural basis of community poverty through their newly radicalized responses to issues of class, race, and gender in urban America.[66]

In March 1971 a number of leading Lords activists arrived in Puerto Rico to promote the island's independence movement as part of an ill-conceived new campaign called the Break the Chains Offensive (Ofensiva Rompe Cadenas). The Lords activists on the island soon found themselves marginal to existing political organizations on the left such as the Partido Independentista Puertorriqueño and the Movimiento Pro-Independencia (later transformed into the Partido Socialista Puertorriqueño).[67] With increasing involvement in the Puerto Rican independence movement and the setting up of new Lords branches on the island, there was a progressive dissipation of energies overseas by the US-born activists. The nationalist sentiments behind Puerto Rican independence began to conflict with the multiethnic character of Young Lords supporters in the North American ghettos and to weaken their ability to form effective alliances with other progressive groups in the city.[68] The Lords' grassroots strength drew on the work of many activists who were not Puerto Rican; with new waves of immigration during the 1970s from other parts of the Caribbean and Central America, the *barrio* was becoming more ethnically diverse at the precise moment that the Lords political agenda was becoming more narrowly nationalistic in its focus. In any case, the Young Lords had always been an authentic voice of the second- and third-generation Puerto Ricans, the *Nuyoricans,* rather than a continuation of the island-based political struggles of the 1930s and 1940s. By 1970 about half of the US Puerto Rican population had been born in the US compared with a quarter in 1950, a demographic shift that lessened the poignancy of demands for Puerto Rican independence. Perhaps the main continuity with the earlier struggles on the island was a fervent anti-Americanism and a resistance to the assimilation of Puerto Rican culture into mainstream American society. The clarity of the independence stance had also been confused by the island's commonwealth status from 1953 onward and the widely boycotted 1968 plebiscite that rejected breaking ties with the US. The shift in focus from local community issues to directly challenging US geopolitical interests in the Caribbean provoked a far more intolerant stance by the American state, fearful of an expansion of Cuban influence. The US government characterized the Lords as a Maoist organization dedicated to the twin goals of destabilizing American society and achieving Puerto Rican independence. With increasing FBI infiltration, internal tensions within the organization were exacerbated and the Lords became embroiled in internal strife, violence, and suspicion.[69]

In the wake of the unsuccessful Rompe Cadenas offensive, the efforts to organize the independence struggle in Puerto Rico were abandoned and the Young Lords decided to devote their energies to the clandestine infiltration of labor unions with a longer-term aim (which was never realized) of resurfacing in the late 1970s. In the summer of 1972 the Lords held their "first and last" national congress in New York City and renamed themselves the Puerto Rican Revolutionary Workers Organization (PRRWO). The relaunched PRRWO sought to build alliances with radical labor groups such as the Revolutionary Union, I Wor Kuen, and the Black Workers Congress, with the ultimate aim of establishing a multiethnic communist party in the United States. The emphasis on street-based recruitment in the early years of Lords activity was to be supplanted by a new emphasis on factory-based organization as a part of a drive to "proletarianize" the organization at a time of quickening deindustrialization and a decline in traditional forms of working-class politics. To underlie this shift, resources were progressively withdrawn from community-based projects and channeled into attempts at workplace organization.[70] Yet this newfound emphasis on class organization proved difficult to reconcile with the largely street-based experiences of Lords supporters. The significance of ethnicity as a cultural resource for the raising of political consciousness became weakened rather than strengthened with the creation of the PRRWO, and the new strategy conflicted with the very limited degree to which Puerto Ricans had organized along class lines in the past. In the 1930s, for example, the presence of an organized Puerto Rican working class in the city had been largely restricted to just two unions: the International Cigar

Makers Union and a Spanish-speaking branch of the International Ladies Garment Workers' Union. By the late 1950s one estimate suggests that no more than 30 percent of Puerto Rican workers were unionized. These low levels of unionization can in part be explained by the persistent complaints filed by Puerto Rican workers against many powerful unions in New York City such as Local 122 of the International Jewelry Workers Union, Local 229 of the AFL Pulp, Paper and Sulphite Workers and Local 1648 of the Retail Clerks Union and the inadequate response of the national AFL-CIO union hierarchy to their grievances.[71] For a combination of historical reasons derived from ethnic and racial divisions within American labor (which have undermined the coherence of the American left throughout the twentieth century), there was little prospect of any effective connection between the street-based campaigns and the fragmentary Puerto Rican organization within the workplace.[72]

In 1970 there were more than a dozen branches of the Young Lords active across the United States, but within three years only four were left. As early as 1970, at the peak of their public profile, tensions emerged at a national level leading to a formal split between Chicago and New York: the Chicago Lords charged the New Yorkers with adopting an increasingly abstract ideological stance, whereas the New York activists became frustrated with what they saw as a lack of national leadership or organizational coherence in Chicago. Members began to drift away, either dropping out of political activity or joining rival organizations such as the Puerto Rican Socialist Party. Other factors also began to sap the strength of the organization at this time: there was a steady loss of young people to Vietnam, drugs, organized crime, and gang violence. In 1973, for example, there were violent clashes in Park Slope, Brooklyn, between the Machetes (an offshoot of the Young Lords) and various Italian-American street gangs.[73] The resurgence of gang violence in the 1970s represented a reversion to the extensive turf-based wars of the 1950s and a decline of youth-based political activism. Increasing social and political fragmentation began to destroy any solid basis for community mobilization. The utopian rhetoric of the 1960s began to lose its allure in the face of the new social and economic realities of industrial decline and worsening poverty. Furthermore, many of the community-based programs and educational projects that emerged from the 1960s became heavily dependent on government funding, leaving them especially vulnerable to the fiscal crisis that swept through American cities in the 1970s. The deindustrialization of New York City saw the Puerto Rican poverty rate spiral from 28 percent in 1970 to 48 percent by 1988. Disparities in the labor market widened as an increasingly informalized and service-based urban economy developed in tandem with the city's emerging role within the new global economy.[74] By the 1980s the Young Lords seemed destined to become a minor footnote in the turmoil of twentieth-century US politics, their political legacy lost forever amid the new political and economic realities of the Reagan era.

4.4 The Power of Memory

The 1990s have witnessed a resurgence of interest in the Young Lords driven by a new wave of political activism that seeks to establish connections between place, environment, and cultural self-determination. The Lords have become part of the fabric of radical political history in New York City through their eloquent exposure of social injustice in the face of widespread indifference and hostility. The Lords owed much to the radical traditions of late nineteenth-century thinkers such as Eugenio María de Hostos and Ramón Emeterio Batances with their emphasis on Antillean liberation. Their political agenda also reflected a degree of continuity with the earlier Puerto Rican organizations of the 1940s and 1950s in terms of community protection against racism, police harassment, and the devastating health effects of slum housing. The Lords rejected the weakly reformist and clientelist strategies of the Democratic Party and sought instead to appeal directly to their own people, inspiring them with glimpses of a different kind of society molded on the needs and aspirations of the ghetto.

In their focus on environmental and public health issues, in conjunction with the high-profile role of women within the organization, the Lords were a forerunner of emerging minority campaigns for environmental justice in the 1980s and 1990s. These observations are significant in two respects. First, much contemporary debate surrounding environmental justice activism pays little attention to the historical evolution of environmental politics. In this instance, a dynamic reinterpretation of Puerto Rican ethnicity played a crucial role in allowing new conceptions and understandings of the Puerto Rican experience to emerge within a broader context of radical social and political change in urban America. Second, the demise of the Young Lords holds important lessons for community-based struggles on the left and their prospects for challenging structural dimensions to poverty and inequality in an international context. We can learn as much from the history of political failure, and its imprint upon collective memory, as from the institutionalized remnants of former victories. Histories of improvements in public health and the urban environment have often placed great emphasis on general advances in nutrition and living standards, rather than focusing on the crucial role of political advocacy in driving progressive change. The radicalization of Puerto Rican politics in New York City revealed the persistence of health inequalities and squalid urban conditions in spite of postwar economic prosperity. The urban environment had now become linked to a much more wide-ranging political agenda than the technical discourses of civil engineering and city planning that dominated urban policy making until the late 1960s. The focus on the human environment transcended dualistic conceptions of nature and culture and opened up new possibilities for progressive environmental politics. The exposure of the structural dimensions to poverty, ill health, and urban blight also challenged the persistence of moralistic and behavioralist dimensions to environmental thought inherited from the nineteenth century.

The Young Lords developed a broad political agenda ranging from poor sanitation in the garbage-strewn *barrio* to a questioning of gender inequalities and access to health care, and in doing so enriched the Spanish-speaking public sphere within the city. They began their protest by focusing on garbage collection as a symbolic form of collective consumption that revealed the social and political marginality of their community. Squalid living conditions in the ghetto were produced and sustained through a nexus of political and economic interests ranging from the disinvestment strategies of slum landlords to the institutionalized racism of government agencies. Their vision of urban society rested on a series of far-reaching social transformations, in contrast to the fragmentary aesthetic interventions associated with urban beautification and the Olmstedian traditions of landscape design (see chapter 2). Although the movement was to fade away in the early 1970s, their campaigns highlighted significant areas of public policy: their health activism raised awareness of the continuing prevalence of tuberculosis in urban ghettos; their focus on lead poisoning formed part of a rapidly emerging political and scientific consensus in the 1970s that contributed toward the founding of the Montefiore Medical Center Lead Poisoning Prevention Project in New York; and in 1971 the issue of childhood nutrition prompted Mayor Lindsay to make the provision of breakfast mandatory in city schools. The Lords' explicit connections between environmental protection and the empowerment of Latino youth would be the inspiration for some of the most high-profile New York environmental justice campaigns in the 1980s and 1990s, such as the Toxic Avengers who sought to protect their Brooklyn neighborhood from toxic waste imports (see chapter 5). Leading activists within the Lords also went on to lead radical environmental organizations in the city such as the South Bronx Clean Air Coalition and El Puente, providing further continuity with the political struggles of the past.[75]

A distinctive and enduring dimension to the political legacy of Puerto Rican environmentalism has been the significance of anticolonial struggles against environmental degradation. A strong pretext for the emergence of demands for Puerto Rican independence by the Young Lords and other radical nationalist groups has been the scale of the devastation on the island wrought by US capital. This international emphasis has fostered links between leading environmental groups such as Misión Industrial in Puerto Rico and the US-based antitoxics movement that developed in the wake of the Love Canal chemical pollution scandal of 1978. Leading Puerto Rican environmentalists such as Neftalí García Martínez have made explicit connections between the island's colonial legacy and the lax enforcement of US environmental law by the Environmental Protection Agency and the Occupational Safety and Health Administration.[76] The challenge to corporate polluters has been consistently seen as part of a wider struggle for greater social and political determination: the anticolonial environmental agenda finds similar processes of disempowerment and impoverishment at work both within the island of Puerto Rico and across the polluted ghettos and minority-inhabited toxic enclaves of North America. The colonial legacy of dismal living conditions has placed the protection of human health at the center of the Puerto Rican environmental movement, as organizations such as Comité Pro-Rescate de Nuestra Salud (CPRNS) (Committee to Rescue Our Health), Misión Industrial, and the feminist organization Organización Puertorriqueña de Mujeres Trabajadoras (OPMT) have led high-profile campaigns against the transformation of their island into a toxic sink.[77] Recent campaigns by the South Bronx Clean Air Coalition, for example, to close a medical waste incinerator involved battling with Browning Ferris Industries, which is simultaneously under attack in Puerto Rico for a proposed giant waste dump within the San Juan watershed. The political dynamics of environmental activism in New York City are inseparably linked to environmental degradation in Puerto Rico, just as emerging demands for environmental justice in Los Angeles owe much to the poisoned *maquiladoras* of the Mexican border.[78] A diasporic environmentalism has grown out of shared grievances and displacements as new alliances are made between disparate voices determined to protect their communities from ill health and ecological devastation.

Through their multidimensional commitment to social change, the Young Lords represent an important yet little-known example of urban social activism. In this respect their significance has more in common with historical examples of urban struggles such as the Paris Commune of 1871 or the Glasgow Rent Strike of 1915 than with the technical elites who pioneered much of the urban and environmental reforms of the twentieth century. Yet their inability to connect between workplace organization and neighborhood protest presents a distinctively American slant on the genesis of urban social movements, where issues of place, community, and identity have played a pivotal role in raising levels of political consciousness.[79] If we examine key interventions in the urban studies literature we find that the process of capitalist urbanization has periodically generated wide-ranging challenges to the inequalities of modern societies. The contradictory dimensions to social and economic change are tightly compressed in urban space as a volatile arena for political action. The urban theorist Manuel Castells emphasizes how urban social movements, despite their immense social and cultural diversity, have three main features in common, all of which we have observed in the case of the Young Lords. The first of these is the demand for more equitable patterns of service provision for collective goods such as sanitation, housing, and health care. A second feature, which contributes to the political salience of collective consumption issues, is the demand for greater cultural and political visibility. And the third is the desire for greater control over urban space as a means to foster a different kind of social structure. These political, cultural, and spatial aspects to urban social movements suggest a very different pattern of organization and mobilization from middle-class neighborhood movements focused on single-issue campaigns.[80] Although the oppositional agenda of these different kinds of protest were sometimes complementary (over urban renewal or opposition to the Vietnam War, for example), the basis for community mobilization was starkly different. A fundamental dimension to the Lords' political strategy was a desire to alter relations between space and society through demands for greater local autonomy and self-determination. In this respect the myriad of concerns were fused into an alternative kind of urbanism, or as Castells puts it "a struggle for a free city, a *citizen movement.*"[81] This reflexive goal, as we saw, also involved a challenge to social relations within the community as well as direct interventions in the social production of space. By occupying strategic spaces within the city, the Lords sought to challenge the dominance of exchange values over use values in the spatial organization of the city. By occupying spaces that were symbolic of the existing social order and changing their functions, they created a situationist city of radical gestures for the delight and consternation of the Puerto Rican community.[82] Ultimately, however, the tension between cultural nationalism and revolutionary ideals became impossible to resolve. The liberation of an imaginary Borinquen held little wider resonance for the diverse and rapidly changing community of the *barrio*. The dynamics of community mobilization began to conflict with an increasingly abstract ideological agenda. As the lived spaces of everyday life featured less prominently in their campaigns the movement began to lose its transformative power and gradually slipped from public view.

5

Rustbelt Ecology

The system works only if waste is produced. . . . Postindustrialism recycles; therefore it needs its waste.

—*Giuliana Bruno*[1]

He got out of the car and climbed an earthen bank. The wind was stiff enough to make his eyes go moist and he looked across a narrow body of water to a terraced elevation on the other side. It was reddish brown, flat-topped, monumental, sunset burning in the heights, and Brian thought he was hallucinating an Arizona butte. But it was real and it was man-made, swept by wheeling gulls, and he knew it could be only one thing—the Fresh Kills landfill on Staten Island.

—*Don DeLillo*[2]

Old, deteriorating, manufacturing districts have become home to the modern city's waste water treatment plants, garbage truck transfer yards, power plants, oil and chemical storage depots, expressways

> and incinerators. Beside them are the homes of African-Americans, Puerto Ricans, Eastern Europeans and other impoverished minorities.
>
> —*Andrew White*[3]

In the previous chapter we saw how marginalized voices in New York's Puerto Rican community articulated a perception of environmental politics rooted in structural demands for social justice that went far beyond the aesthetic transformation of urban space. The 1960s marked a pivotal transition between the "expert vision" inherited from the Progressive era and emerging grassroots demands for more inclusive forms of urban planning. In this chapter we develop the theme of environmental justice through an examination of land use conflict in the 1990s. We examine environmental politics in the Greenpoint-Williamsburg district of Brooklyn, which forms part of the American rustbelt of abandoned industrial spaces (figure 5.1). The emergence of the rustbelt across the industrial states of the North and East has been one of the most significant spatial dimensions to the transformation of the US economy since the mid-1960s, as the fulcrum of economic growth has moved toward the rapidly growing sunbelt states of the South and West. Former centers of manufacturing industry such as Greenpoint-Williamsburg have found themselves marginalized by a combination of technological, social, and political developments, which have favored the creation of new kinds of production complexes often underpinned by powerful levels of federal support. The emergence of "new industrial spaces" has become the focus of an extensive literature, but much less attention has been devoted to the physical transformation of former sites of industrial production. We know little about the political and economic dynamics behind the development of such "toxic industrial spaces" and the environmental inequalities associated with new patterns of economic restructuring.[4]

Some indication of the severity of urban decline in Greenpoint-Williamsburg is suggested by the loss of over 20 percent of its population between

5.1 Looking north from the Williamsburg Bridge in 1981.
Source: Courtesy of Al Sacco. City Limits photograph collection.

1970 and 1980. Since the early 1970s this blighted urban landscape has seen spiraling levels of poverty, violence, and interethnic conflict. Yet we should be wary, warns the urban scholar Rosalyn Deutsche, of teleological representations of urban decay that risk the naturalization of urban poverty and the xenophobic perpetuation of "New York's sociospatial hierarchy."[5] The recent experience of Greenpoint-Williamsburg points to possibilities for transforming the politics of pollution in impoverished urban areas. In the 1990s this deprived urban backwater has become emblematic of a new kind of political self-confidence on the part of ethnic ghettos long demonized in the popular imagination. The Brooklyn waterfront has become the focus for new forms of environmental activism with potentially far-reaching regional implications for the distribution of unwanted land uses.

The focal point for the resurgence of radical grassroots campaigns on behalf of the social environment has been the unresolved dilemma of how to get rid of the growing quantities of waste produced in late twentieth-century America. Places that had been integral to the production process in the Fordist era have been transformed into "end cities," as waste-handling companies colonize empty pockets of land zoned for industrial purposes. The processing and disposal of waste products presents us with one of the sharpest geographical indices of social power etched into the urban landscape. A waste-based economy has emerged in Greenpoint-Williamsburg as the symbiotic inverse to the glittering skyline of Manhattan.[6] In the wake of New York's industrial decline, the production of waste has become one of the most significant material outputs of the metropolitan region. By the early 1990s, for example, waste paper had become the main export commodity from the city following the expansion of the so-called FIRE (finance, insurance, real estate) economy in the 1980s.[7] The city generates nearly 14 million tons of solid waste a year, of which around 40 percent is derived from municipal wastes (residential, city agency, and institutional sources) handled by the city's Department of Sanitation while the other 60 percent is composed of commercial wastes disposed of by private carters.[8]

New York has never found a satisfactory way of handling its waste. All the main options for dealing with the city's waste present an array of difficulties: both landfill and incineration are unpopular and environmentally damaging; the recycling of garbage is expensive, complex, and limited in its potential contribution; and the reduction of waste at source implies a fundamental alteration in patterns of production and consumption beyond the scope of municipal governance.[9] Disputes surrounding the human and environmental effects of the waste economy raise perplexing issues about the nature of expanded production and consumption that cannot be addressed by ad hoc policy interventions or by facile appeals for behavioral change. The production of waste has increased over the last thirty years in spite of greater environmental awareness and a plethora of recycling initiatives.[10] The proliferation of waste and its ideological alter ego of recycling have become characteristic of the consumption-orientated post-Fordist city with its diverging realms of social and economic activity.

When we examine the contemporary resurgence of recycling activities in Western cities, we find a curious degree of congruence between premodern and postmodern environmental formulations. The perceived incompatibility between the modern metropolis and environmental sustainability is rooted in a long-standing suspicion toward the circulatory dynamics of capitalist urbanization and the modernization of urban infrastructures to serve metropolitan needs. Leading commentators on nineteenth-century urbanism such as Friedrich Engels and Victor Hugo feared the deleterious consequences of allowing human wastes to be flushed into newly constructed sewer systems, with the consequent loss of nitrogen for agriculture (see chapter 1). In late nineteenth-century New York, for example, abattoir wastes and other organic materials were taken to Barren Island to be boiled down into grease, oils, and fertilizers. A wide range of recycling activities were undertaken for economic reasons, providing marginal livelihoods for thousands of New Yorkers working as ragpickers, scavengers, and "scow trimmers."[11] Yet the economic viability of these activities has always been highly marginal on account of fluctuations in the value of secondary materials and longer-term shifts in the nature and composition of urban wastes. "We have not yet reached any very satisfactory knowledge as to the conversion of waste into wealth," noted New York's leading sanitation pioneer, George Waring, in 1895: "While the theoretical value of discarded matters is recognized, the cost of recovery is still an obstacle to its profitable development."[12] Waring's observation remains crucial to any understanding of the city's contemporary waste dilemma, since recycling has emerged as a panacea for many environmentalists who oppose capital-intensive forms of waste disposal such as landfill and incineration. But the promotion of recycling for primarily environmental reasons since the 1970s has not provided an economically viable alternative to the city's continuing reliance on landfill for the disposal of its waste. Recycling has become a symbol of the future without any clearly defined role in tackling the underlying dynamics of waste production. And radical commentators on modern waste policy such as Stephen Horton and Simone Martzloff have actually opposed postconsumer recycling altogether as a diversion from more fundamental political and economic choices.[13]

The politics of waste in New York City has been dominated since the early 1970s by one principal dilemma: whether the city's vast Fresh Kills landfill site on Staten Island can be replaced by a new generation of incineration plants to handle the city's municipal waste stream.[14] This landfill began operation in 1948, having been identified as a suitable site for land reclamation by Robert Moses in the 1930s. Over time Fresh Kills gradually emerged as the dominant waste disposal facility for the city and is now the largest garbage dump in the world. Its vast mountains of rubbish reach 500 feet high—dwarfing the pyramids of Egypt—and the 2,100-acre site is one of the most significant human artifacts observable from space.[15] The level of local opposition to Fresh Kills by Staten Island residents reached such proportions that a referendum was passed in 1992 to secede from New York City and become an independent municipality within New York State. The potential for independence from the disposal of New York's waste is complicated, however, by the island's dependence on the city's water supply system. In any case, the city is committed to closing the Fresh Kills landfill by the end of 2001, which has necessitated a fundamental rethinking of the city's waste strategy.[16] The closure of the site is in itself a major undertaking, involving everything from the long-term control of methane emissions to the extensive landscaping of the site, at a projected cost of over $800 million. In response to this looming garbage crisis, the city has, since the late 1970s, proposed the construction of a new generation of giant waste incinerators, the first of which was proposed for the derelict Brooklyn Navy Yard in the heart of Greenpoint-Williamsburg. This plan provoked a remarkable transformation in the dynamics of grassroots political activism. New kinds of alliances emerged across Greenpoint-Williamsburg in response to a common threat represented by a major new source of air pollution in an already heavily polluted and deprived urban neighborhood. The eventual defeat of this proposal, against all odds, reveals an important shift in the nature of urban policy making since the Moses era of infrastructure development. A new organizational sophistication on the part of grassroots activists in combination with the ethical zeal of the civil rights legacy has had far-reaching consequences for environmental politics in the city.

5.1 Across the Great Divide

On 14 January 1993 a crowd of 1,500 mainly Latino and Hasidic community activists and their supporters marched across the Williamsburg Bridge from Brooklyn to Manhattan to protest the proposed building of a giant waste incinerator on the derelict Brooklyn Navy Yard site in the midst of one of the poorest communities in New York City. This striking display of racial unity between the Latino and Hasidic communities, which had for many years been locked in bitter conflict, attracted widespread media interest (figures 5.2 and 5.3). Not only did the march provide visible evidence of an unprecedented degree of interethnic reconciliation but it powerfully underlined the webs of power that link the creation of toxic industrial spaces to the political and economic elites of Manhattan. In effect, the march sought to redress spatial inequalities in urban pollution underpinned by the evolution of twentieth-century zoning law and the legacy of technocratic decision making.[17]

The Greenpoint-Williamsburg district of Brooklyn has one of the most complex ethnic structures in the city, comprising Latinos, Hasidic Jews, and an array of smaller ethnicities such as Polish and Italian Americans. The Latino community is dominated by Puerto Ricans, who arrived in large numbers in the 1940s and 1950s and were later joined by Dominicans, Mexicans, and other Spanish-speaking immigrants, making Latinos the largest ethnic group in the neighborhood. The Latinos are separated from the Hasidic Jews by the aptly named Division Avenue, which runs through the heart of Williamsburg from the

5.2 Dioxin death mask ritual? The march across the Williamsburg Bridge to protest against the Brooklyn Navy Yard waste incinerator, January 1993.
Source: Courtesy of Debra DiPeso. City Limits photograph collection.

5.3 The march across the Williamsburg Bridge to protest against the Brooklyn Navy Yard waste incinerator, January 1993.
Source: Courtesy of Debra DiPeso. City Limits photograph collection.

East River.[18] The Jewish community is dominated by the Satmar Hasidim, who arrived in the wake of the Holocaust and continue to draw immigrants from eastern Europe. Long-standing problems of racism and anti-Semitism in Greenpoint-Williamsburg have been fueled by poverty, unemployment, and fierce competition over access to resources such as housing and education. The urban renewal programs of the 1960s and 1970s contributed to community tensions as vast areas of old apartments were cleared to make way for new high-rise buildings. In 1974, for example, some 500 members of the Latino community occupied the newly built, federally subsidized Roberto Clemente Houses in protest at what they saw as an unfair housing allocation in favor of the Hasidic community. In 1989 the Southside Fair Housing Committee won a settlement that set aside 190 apartments in three public housing projects for Latinos and African Americans, who successfully argued that they had been discriminated against. In 1992 there was renewed conflict over housing allocation between the Latino and Hasidic communities, and many issues of common concern remain unresolved. In addition to struggles over access to resources, there have been sporadic vigilante-led beatings and outbreaks of violence. Against this background, the successful community-wide mobilization against environmental pollution takes on a special significance.[19]

We can trace the origins of this community reconciliation to early 1991 when Luis Garden-Acosta, a former Young Lords activist and chief executive of the Latino educational foundation El Puente, met with Rabbi David Niederman, director of the United Jewish Organizations (UJO), a Hasidic social support group, in order to work together in what Randy Shaw describes, in his review of political activism in 1990s America, as an "astonishing example of tactical coalition building at the grassroots level."[20] Garden-Acosta described his initial approaches to the Satmar Hasidim as "like Nixon going to China," but a common program of environmental concerns was quickly identified. The Hasidic community living in close proximity to the Brooklyn Navy Yard already recognized that they would be unable to defeat the incineration proposal on their own, since an earlier Satmar-led march in 1985 had failed to prevent the city from signing a contract to proceed with the incineration plant. Early contact between Satmar and Latino activists had been fostered by Latino youth groups such as the Toxic Avengers who began to identify a common agenda. A joint Latino-Hasidic demonstration was organized against a hazardous waste storage facility run by the multimillion-dollar Radiac Corporation, which handles toxic, flammable, and radioactive wastes in the middle of Greenpoint-Williamsburg just a few meters away from a school.[21] The success of early efforts at political partnership with the Latino community began to alter the scope of environmental politics. The pivotal moment, however, which created the momentum for the march across the Williamsburg Bridge was the creation of a multiethnic forum called the Community Alliance for the Environment at a public meeting held in the autumn of 1992:

> On the eve of the 1992 Democratic Convention, the pair [Garden-Acosta and Niederman] brought together more than a thousand people—not only Latinos and Hasidim, but also African-Americans, Polish- and Italian-Americans—in Williamsburg's first environmental town meeting. Garden-Acosta asked participants to stand if they were ready to fight the incinerator and to repair the division between ethnic groups. A thousand people rose from their seats, and the Community Alliance for the Environment (CAFE) was formed.[22]

Since the early 1990s Greenpoint-Williamsburg has emerged as a focal point for new demands for environmental justice that have succeeded in altering the city's environmental policy making in a series of complex ways. The area has the highest proportion of land devoted to industrial activity of any community district in the city, with levels of pollution almost 60 times the national average for residential neighborhoods (figure 5.4). Greenpoint-Williamsburg is completely encircled by health-threatening facilities, including the Brooklyn Navy Yard site, petrochemical plants, a transfer station for low-level radioactive waste, a major sewage works (the much-maligned Newton Creek plant), the Brooklyn-Queens Expressway, widespread lead contamination, commercial wastes trucked from several dozen private transfer stations, a massive underground oil spill, and the recently closed Greenpoint waste incinerator (figure 5.5).[23] In the nearby housing

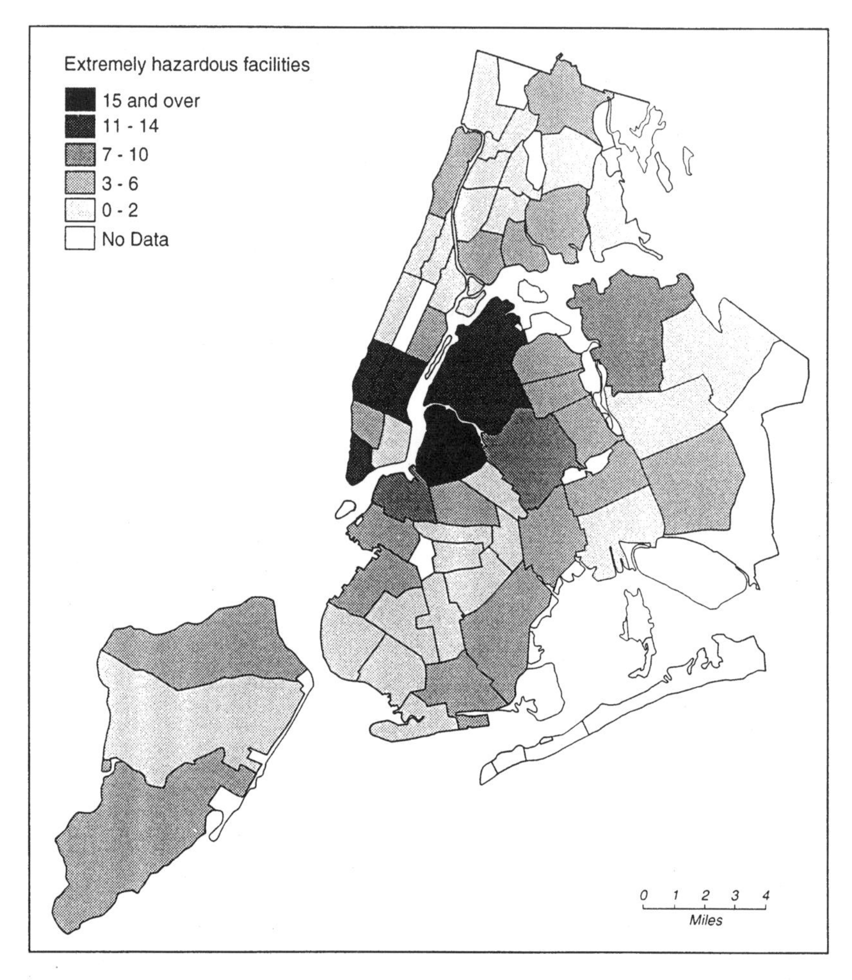

5.4 Variations in the distribution of toxic facilities across the 59 New York City community districts. The heaviest concentration marks Greenpoint-Williamsburg.
Source: City of New York Department of City Planning.

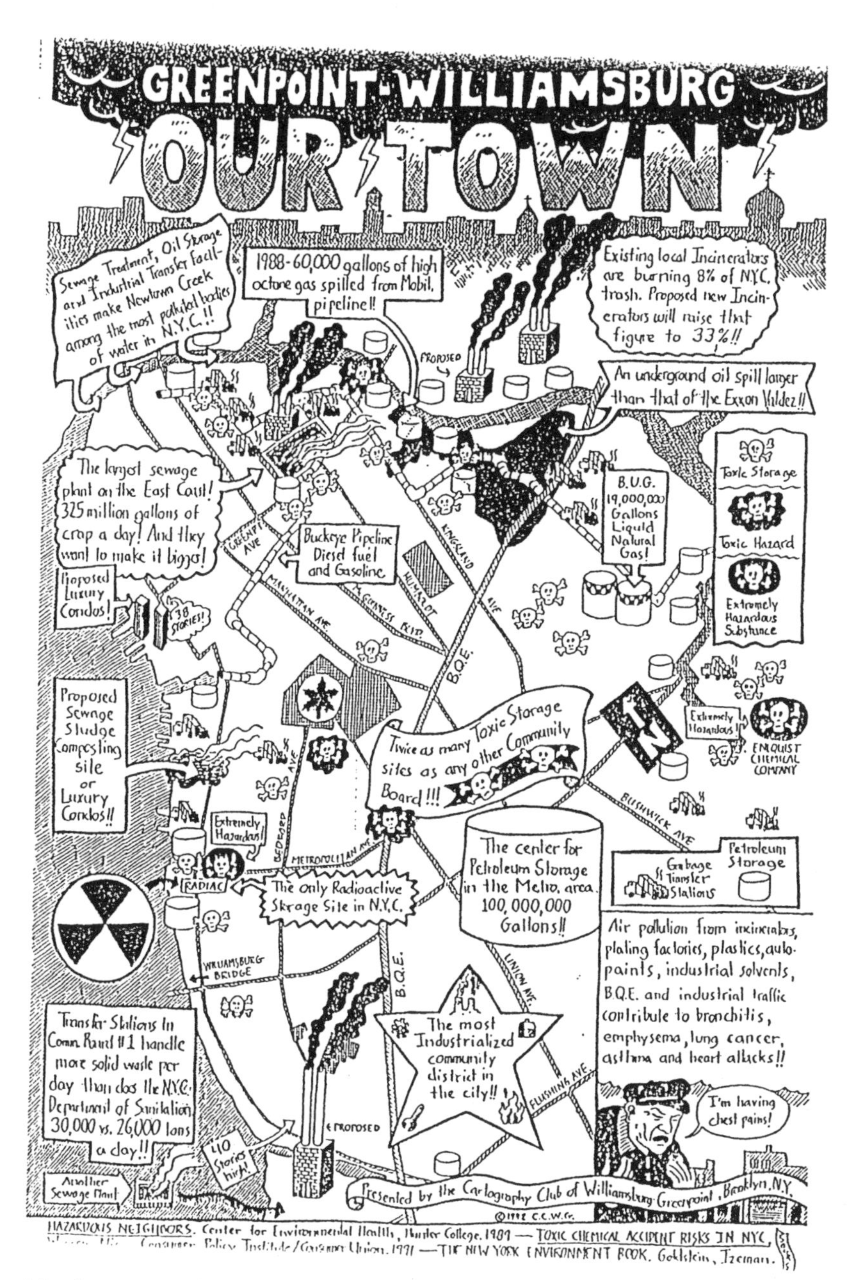

5.5 A toxic map of Greenpoint-Williamsburg.
Source: Cartography Club of Williamsburg-Greenpoint, Brooklyn.

projects of Fort Greene, for example, a study found that 26 out of every 1,000 children were admitted to hospital for bronchitis or asthma in one year alone, which is significantly higher than the Brooklyn average.[24] And these rates of emergency hospitalizations are exacerbated by inadequate primary health care and the widespread lack of adequate medical insurance. More specifically, research has shown that rates of asthma morbidity and mortality in New York City are not only substantially higher than the national average but that there is a higher prevalence in poorer neighborhoods within the city. The incidence of the disease is also disproportionately concentrated among nonwhites, with hospitalization and death rates among blacks and Latinos up to five times higher than those of whites.[25] When we add to this picture the results of recent research into the reemergence of tuberculosis and other preventable diseases, we find that the intersection of race, class, and place has produced a landscape of health inequalities with far-reaching political implications.[26] Growing public recognition of these health inequalities in the 1990s has radicalized debate over land use planning and its connection with issues of social and environmental justice.

5.2 Pollution and the Politics of Resistance

> It would be hard to find a more detested . . . proposal in New York City over the last 20 years than the Brooklyn Navy Yard incinerator.
>
> —*Eric Goldstein*[27]

The most important catalyst for the emergence of radical environmental activism in Greenpoint-Williamsburg has been the proposed construction of a giant waste incinerator on the site of the former Brooklyn Navy Yard. During the Second World War this site held the largest naval construction facility in the United States, employing over seventy thousand workers. By 1965, however, fewer than seven thousand people worked there, and the 255-acre site was closed the following year. The closure of the Brooklyn Navy Yard, once the biggest employer in the region, is a poignant symbol of the city's postwar industrial decline.[28] In the

1990s the Brooklyn Navy Yard had again become a symbol of urban change, in this instance as a focus for political resistance against pollution.

The background to the proposed construction of a new waste incinerator at the Brooklyn Navy Yard site has been the city's attempt to plan for the future of its waste disposal after the closure of the Fresh Kills landfill on Staten Island. The growing reliance on Fresh Kills contrasts with the variety of earlier options, which once included ocean dumping, a variety of recycling initiatives, over 80 landfill sites (in the early 1930s), some 22 incineration plants, and thousands of smaller apartment house incinerators.[29] The postwar contraction of waste disposal options can be attributed to a series of developments. First, there have been increasingly restrictive environmental regulations, marked in the 1930s by the ending of ocean dumping (following court action against the city by the neighboring state of New Jersey), and in the 1970s by ever more stringent federal controls on the use of landfill and incineration. Second, there has been a progressive exhaustion of landfill capacity in and around the city (a development that has been accelerated by the legal protection of wetland ecosystems). And in the postwar era there have been a set of interrelated socioeconomic changes associated with the rise of mass consumption, the rapid growth of the municipal waste stream, escalating labor costs in waste management, and declining markets for many recycled materials.[30] In the early 1980s the city's Sanitation Department concluded that the potential contribution of recycling and incineration had been misjudged and consequently looked to an upgrading and extension of operations at the Fresh Kills landfill as the only viable option. Yet the detailed engineering analyses carried out revealed "the frightening dimensions of the waste disposal problem" for the first time. Not only was Fresh Kills found to be approaching full capacity, but the site had been in violation of state law since 1980 for the release of millions of gallons of toxic leachate into nearby watercourses.[31]

Recent efforts to avert the city's dependence on Fresh Kills by an expansion of incineration date from Mayor Lindsay's administration in the early 1970s, but the city's fiscal difficulties at that time put an abrupt end to any program of capital construction. In the late 1970s, however, in the wake of the OPEC-induced energy crisis, President Carter introduced measures to subsidize nonfossil fuel sources of energy and ensure that utilities purchase energy from alternative sources. This shift in energy policy received strong backing from the packaging lobby, investment bankers, and power plant construction firms. These interest groups were nervous that recycling and waste reduction might emerge as the main alternative to landfill as a result of the politicization of waste issues and the passing of mandatory recycling and returnable container legislation in a number of American states. Incineration plant constructors also included many engineering firms formerly involved in the nuclear industry (examples include Westinghouse, Combustion Engineering, Babcock & Wilcox, and General Electric) which was in steady decline in the US, particularly in the wake of the Three Mile Island nuclear accident of 1979.[32] By the 1990s over two-thirds of incineration plant construction was dominated by three firms—Wheelabrator, American ReFuel, and Ogden Martin—which were themselves subsidiaries or partners of the two most powerful waste handling corporations Waste Management and Browning Ferris Industries (now part of Allied Waste).[33] Thus waste management had become dominated by a powerful nexus of commercial interests who have worked assiduously to build an international ascendancy for the provision of municipal services, in an analogous fashion to other recent developments in the fields of water, food, and information technology. The power of these waste management empires has grown as federal sources of research and expertise have been scaled down since the 1970s, leaving municipal governments increasingly reliant on the waste industry itself for policy guidance and technical expertise. At a political level, extensive contacts and campaign donations have fostered close links between the waste industry and municipal governments. This symbiotic relationship has developed in an era of fiscal retrenchment and the decline of municipal managerialism as an ethos for the provision of environmental services (see chapter 1). The extensive use of municipal bond finance for new public infrastructure projects serves to confine much of the economic risk within the public sector while ensuring long-term profits for private-sector operators. The high point of the incineration renaissance was reached in the early 1990s, when incinerators handled around 16 percent of the US municipal waste stream, but there were already signs of political, economic, and structural weaknesses running through this capital-intensive approach to urban waste disposal.

In 1978 Mayor Edward I. Koch proposed the construction of the first of a new generation of incineration plants at the Brooklyn Navy Yard with the strong backing of business interests, key media outlets, and powerful construction unions. A Citizens' Advisory Committee was set up in 1981 in order to streamline public involvement in the planning process, with one-third of its members appointed by the city's Department of Sanitation, one-third by the Brooklyn borough president, and one-third by community boards near the proposed site. By 1985 an environmental impact statement had been completed and the city had signed an initial contract for $290 million, but construction of the plant began to face a series of unexpected political and legal challenges.[34] The Citizens' Advisory Committee failed to garner public support for the project because it was perceived to be approving a decision that had already been taken. Although the committee succeeded in pushing through various technical modifications to the project, the broader rationale of providing public legitimation for capital-intensive waste management was a failure. As discussions over the proposed plant moved into the public domain, a sharp polarization of views emerged. The city and its corporate backers found themselves up against a growing array of protests from the Clinton Hill-Fort Greene Coalition for Clean Air, the Consumer Policy Institute, the Environmental Action Coalition, the Lower East Side Coalition for a Healthy Environment (with backing from the United Jewish Council and the Physicians for Social Responsibility), the League of Women Voters, and many others. These different organizations represented the interests of the scientific and medical establishment, various resident and tenant associations, women's groups, environmentalists, and grassroots political activists, and reveal a similarity to other anti-incineration campaigns in Baltimore, Boston, Nashville, and San Diego.[35]

The radicalization of waste politics was bolstered in New York by the disintegration of the political coalition behind the Koch administration which had governed the city since the late 1970s. With the defeat of Koch by David Dinkins in the Democratic primary of 1989 there was much greater openness to minority and environmentalist demands in city government.[36] The left-leaning Dinkins campaign provided a forum for a generation of community activists who had shaped their agendas during the political struggles of the 1960s and 1970s. By 1989 the expansion of incineration had become a prominent issue in the mayoral elections, and Dinkins clinched a narrow victory for the Democrats running on an anti-incineration platform with public endorsements from Brooklyn-based community groups and the city's environmental lobby.[37] The political conflict over the proposed plant intensified in 1991, however, with the abrupt reversal of Dinkins's stance to favor the expansion of incineration, which provoked consternation among his erstwhile political allies. A joint press conference held by the United Jewish Organizations, the New York Public Interest Research Group, and El Puente condemned the city's continuing allocation of noxious facilities to poor minority neighborhoods, with lasting significance for the environmental credibility of the Dinkins administration. The anti-incineration campaign was able to accuse the city of environmental racism, since over two-thirds of the people living within a 4-kilometer radius of the Brooklyn Navy Yard site were classified as minority groups in the 1990 census.[38] By the early 1990s the issue of environmental racism had become an integral dimension to debates over land use policy and environmental planning: it was no longer possible to conceive of technical aspects to planning procedures as socially or spatially neutral without risking public opprobrium.

An important issue for mobilizing public concern is the fear that large-scale incineration will threaten human health. The emission of chlorine-containing compounds such as dioxins is at the center of ongoing epidemiological disputes about the possible effects of estrogen-mimicking "environmental hormones" on human immunological and reproductive organs. The known carcinogenic properties of dioxins have already gained notoriety through the Seveso chemical poisoning in Italy in 1976 and the use of the Agent Orange defoliant in Vietnam.[39] The efforts of a leading environmentalist, Barry Commoner, have been pivotal in promoting scientific evidence against the safety of dioxins. Commoner, in advance of the emerging consensus against chlorine, succeeded in politicizing the science of waste incineration to an unprecedented extent and enabled community activists to utilize the latest advances in international toxicology and public health research.[40] Other pollutants associated with incineration include heavy metals; especially worrisome was the projected release by the proposed incinerator of some 15 tons of lead a year. Existing lead levels in New York are already a cause for concern: one report revealed that 29 percent of nearly 5,000 children screened in the summer of 1991 had blood lead levels exceeding the official 1978 standards of no more than 30 micrograms per deciliter in children.[41] The city had in fact attempted to block the construction of an incinerator in neighboring New Jersey on the grounds that emissions posed a health threat to New York, contradicting incinerator safety claims made for the Brooklyn Navy Yard project. This is an argument that became even more contentious in the late 1990s with the diversion of waste from Fresh Kills to be incinerated in the poor and minority-dominated city of Newark, New Jersey.[42] A further reason for the delay in construction of the Brooklyn plant was uncertainty over the disposal of residual ash from the incineration process, constituting as much as 15 percent of the volume of the original wastes delivered to the plant. A 1987 survey commissioned by New York State found that levels of lead and cadmium in ash from six existing incinerators exceeded federal guidelines and rendered the ash hazardous waste. The lack of any satisfactory arrangements for the disposal of over 900 tons a day of residual ash from the proposed plant led to the initial refusal by New York State to issue a construction permit in 1988.[43] There was also continuing uncertainty surrounding the cost of a major expansion of incineration. Studies commissioned by environmentalist groups suggested that the operation of the plant might be more expensive than anticipated because of debt servicing, technical difficulties, fluctuations in the revenue gained from power generation, and the cost implications of disposing of incinerator ash as hazardous waste.[44] Other environmental concerns included the potential contribution to global warming caused by the release of carbon dioxide and other pollutants (though the release of methane from landfill sites is more significant in this regard)—a locally telling argument because New York City is particularly vulnerable to global warming through the disruption of drinking water supplies by algal blooms in upstate reservoirs and salt water incursion of the Hudson River, in addition to the potential impact of higher sea levels and greater storm intensity on low-lying parts of the city.[45]

Environmentalist groups also opposed the expansion of incineration on the grounds that it would undermine the impetus for recycling and waste reduction in the city.[46] It is doubtful, however, whether recycling and waste reduction could provide an economically viable alternative to the city's continuing reliance on landfill. Estimates in the late 1990s put the cost of recycling at significantly higher than the use of out-of-state landfills after the closure of Fresh Kills. Quite apart from the logistical complexities of operating multimaterial recycling schemes for high-density urban areas, there are a range of economic uncertainties affecting long-term expenditure on labor-intensive forms of waste sorting and the reliability of markets for recycled materials.[47] Although the city was legally mandated in 1989 to achieve a 25 percent recycling rate by 1994, this target has proved elusive. In 1993 the rate of recycling for municipal waste reached around 15 percent but has subsequently slipped, largely as a result of budget cuts affecting the number of collection rounds and a scaling back of educational outreach initiatives. Fluctuations in the price of waste paper have added to the city's difficulties: in 1995 the city was earning around $160 a ton for it, compared with a loss of $5 ton in 1996. With the expansion of the city's annual export trade of 300,000 tons of waste paper in doubt, the entire recycling program has been left in an economically precarious position. This has led to calls from the Giuliani administration to alter the definition of recycling to include abandoned cars and construction waste as the only cost-effective means to reach the city's target.[48]

Historical analysis of the city's waste stream shows that recycling has been primarily pursued for economic rather than environmental reasons. The viability of recycling has declined since the late nineteenth century in response to a range of technological, economic, and medical developments. Organic fertilizers were replaced by the development of chemical fertilizers and new petrochemical products. New paper-making technologies from the 1890s onward brought a collapse in prices for waste paper markets. And food collections for pig swill rapidly declined in the 1950s in order to prevent the spread of trichinosis, so that by the early 1960s only 4 percent of American food wastes were fed to animals.[49] An important study published by the US Environmental Protection Agency in 1972 found that levels of recycling had sharply fallen after World War II.[50] This decline was attributed to the difficulties faced by the secondary materials industry, especially for paper, textiles, and rubber, where small-scale, often family-run enterprises found it increasingly difficult to compete with virgin materials because of rapid technological developments in the capital-intensive extractive industries for the processing of raw materials. As virgin materials improved steadily in their quality and price competitiveness, the waste stream itself began to include increasing quantities of materials such as plastics that were difficult to reuse. The US paper recycling rate fell from 35 percent in 1944 to around 17 percent by 1973, and similar declines are recorded for rubber, aluminum, copper, lead, and other scrap metals.

The contemporary emphasis on recycling as a response to the urban waste crisis is best perceived as an ideological dimension to the post-Fordist waste economy. As Stephen Horton suggests in his analysis of waste politics in Los Angeles, "Recycling, regardless of its limitations, is acclaimed as an emerging alternative, that will soon reduce the need for landfills. Between the threat of the present and the promise of the future, the practice of the past is reproduced."[51] There is, in other words, an emerging synergy between the discourse of a "waste crisis" on the one hand (which may necessitate the expansion of profitable capital-intensive forms of waste management) and environmentalist demands on the other hand to recycle ever greater proportions of a growing waste stream. When the locational dynamics of increasing opposition to landfill, incineration, or waste transfer facilities are added into the equation, it becomes clear that the future direction of urban waste management is highly uncertain. There is local opposition to the presence of any waste-handling facilities in New York, even garages and depots, because of the likely impact on local property prices, and this has tended to reinforce the trend to locate unwanted facilities in industrial or blighted areas, usually in low-income parts of the city with a high proportion of public housing. "Invisible urbanism" has become the latest stage in the fetishization of the urban landscape. "The products must still be made, the power generated, the water pumped, and the sewage treated," notes Thomas H. Garver, "but let it be done elsewhere, and invisibly."[52] The elimination of all vestiges of production from the urban landscape is leading to widening environmental disparities at a regional level, as increasingly competitive forms of local antipollution activism simply redistribute unwanted land uses along an existing axis of property values and neighborhood militancy.

Beyond the epidemiological debates surrounding the safety of incineration emissions, there is a sense in which environmental activism serves as a symbolic form of community empowerment by linking concerns with health, housing, and inadequate political representation. High levels of alienation and skepticism among the city's poor have led to an erosion of trust between municipal government and civil society. Given the failure of the city to adequately look after nearby public housing projects, for example, tenants are naturally doubtful that any new plant would be safely operated. The level of opposition to the proposed Brooklyn Navy Yard incinerator has undoubtedly been intensified by the widespread prevalence of asthma and lead poisoning in Greenpoint-Williamsburg and a sense that poor communities in the city have shouldered a disproportionate burden of environmental hazards: "The Navy Yard plant has become a focal point for the community's simmering resentment about a whole range of issues: pollution from cars and trucks, garbage on the streets, drug abuse, joblessness, inadequate health care and deteriorating housing. It is environmental activism in the fullest sense of the term."[53] There are striking parallels between the polluted Latino communities of Brooklyn and the experience of poor communities on the US-Mexican border: in both cases the lack of political and economic power has led to an environmentally threatening environment. Opponents of the incineration plant depicted Mayor Koch's former Sanitation Commissioner, Norman Steisel, as a kind of Robert Moses figure making deals beyond political scrutiny and insensitive to public concerns.[54] The opposition to the incinerator also pitted the people of Greenpoint-Williamsburg against powerful Wall Street bond underwriters who stood to benefit from the project, as revealed in Randy Shaw's analysis:

> Among Wall Street firms with a stake in the incinerator was Lazard Frères, whose Felix Rohatyn was New York City's most powerful and influential investment banker. Rohatyn's power over municipal credit and city budgets made ambitious elected officials extremely reluctant to oppose him. Wall Street firms were major sources of

> campaign funding for local and state politicians, who could then support projects like the incinerator without having received funds from the project sponsor. . . . The *New York Times* took a vehemently pro-incinerator editorial stance, joining various other power brokers in the city. The law firm retained by Wheelabrator, the company hoping to build the incinerator, made major contributions to local and state legislative campaigns, including $45,000 for Rudolph Giuliani's 1993 mayoral race.[55]

The city's Department of Sanitation has tried to allay public fears of incineration by putting the health and environmental risks in a wider perspective. For example, a survey of environmental pollution in the New York-New Jersey region by the US Environmental Protection Agency found that municipal waste incinerators rank as a very low risk in comparison with the medium risk posed by landfill sites and the very high risk from motor vehicle exhaust fumes.[56] The projected lead emissions from the proposed Brooklyn Navy Yard plant have been singled out as a threat to children's health, yet on one calculation the energy generated by the plant would displace one million barrels of oil a year used by Con Edison, which would emit some seven times more lead for an equivalent generation of energy.[57] Despite the city's efforts to reassure local communities of the safety of incineration, new scientific evidence on the impact of dioxins and other pollutants since the late 1980s has served to heighten rather than diminish public concerns. Recognizing that the political argument over incineration had been lost, Mayor Dinkins quietly dropped the incineration issue by delaying the start of construction from 1994 until 1996 in order to remove the issue from his reelection bid.[58] In the event, Dinkins narrowly lost the 1993 mayoral race in part due to his loss of support from the city's environmental movement and high levels of abstention among nonwhite voters.

By the mid-1990s Mayor Giuliani had also distanced himself from the proposed Brooklyn incinerator and looked to a series of short-term measures to solve the city's waste crisis. In addition to the political death of incineration in New York, wider economic factors have played a role in this shift, such as the phasing out of the mandatory purchase of high-priced electricity from waste-to-energy plants (a legislative relic of the energy crisis in the 1970s) and the development of a new generation of huge regional landfills in Virginia, Pennsylvania, and elsewhere. The newest landfill sites bear little resemblance to the thousands of unlined and leaking municipal dumps that were forced to close after new federal environmental regulations in the 1970s: to some extent the case for both recycling and incineration has relied on a misconception of the latest advances in landfill technologies such as lining technology and leachate control. By the mid-1990s the cost difference between landfill and incineration had grown progressively wider, particularly in the Northeast where contractual controls over waste allowed costs to rise to over $100 a ton for some incineration plants, compared with average figures of around $30 a ton for landfill. In 1997 a ruling by the US Supreme Court against the preferential treatment of incinerators for municipal waste ended the practice of flow control in which waste would be handled by a designated facility. This led in turn to a downgrading in the investment grade for municipal bonds used for waste-to-energy facilities.[59]

In the final instance, the political case for the Brooklyn incinerator was to disintegrate for economic reasons: the real success of the grassroots campaign lay in delaying its construction until the project could no longer be perceived as financially viable. In contrast, the political battle to prevent the construction of the loss-making Newark incinerator in New Jersey took place before the economic arguments could have worked in the community's favor. The local Portuguese community of Ironbound must now endure increasing quantities of New York's waste as the incinerator operators struggle to find sufficient garbage to run the plant. This waste-to-energy facility, now operated by American ReFuel, has emerged as a key element in the city's post-Fresh Kills waste disposal strategy.[60]

Opposition to waste incineration in New York is a far more complex issue than demands for urban air quality. The campaign to alter the city's waste policy has exposed both the intractable dynamics behind the production of waste and also the politicization of place in order to advance an agenda of environmental justice for some of the city's poorest citizens. The fact that a community-based campaign could challenge a significant dimension to urban policy making in the face of virtually every powerful interest group in the city is testament to a shift in the political dynamics of land use planning since the Moses era. In effect, the city's ecological frontier has been extended and redefined in new ways. We saw in chapter 1 how the extension of water technologies into upstate New York linked the city into a vast hydrological system covering thousands of square miles. Similarly, in the case of waste, the impact of the city now extends via long-distance waste transfer by sea to the landfills of the eastern seaboard stretching from New Hampshire to Virginia.

As late as the mid-1990s no municipally collected waste had been disposed of outside the city since the early 1930s, and even the possibility of sending waste from New York City to landfills elsewhere aroused strong passions: in 1989, for example, the town of Benton in Arkansas rejected a contract to receive waste from New York amid public outcry over the threat of syringes from unmonitored medical wastes, and the then-governor of Arkansas, Bill Clinton, supported legislative moves to ban the receipt of out-of-state garbage.[61] From 1997 onward, however, the city began to export wastes outside the New York metropolitan region for incineration in New Jersey and landfill in Ohio, Pennsylvania, and Virginia. This has opened up an entirely new set of political tensions between the city and its emerging environmental sink located in the rural South and impoverished pockets of the East Coast rustbelt. By 1999, for example, Mayor Giuliani and Governor James S. Gilmore III of Virginia were trading insults over the interstate transfer of garbage: to widespread incredulity Giuliani suggested that the city's waste was a fair exchange for the city's cultural and economic contributions to national life. The regional flows of tourists and garbage could be interpreted as a postmodern equivalent of earlier exchanges in food, timber, and other raw materials that built the nineteenth-century metropolis.[62] At a local level, however, the issues are far less clear. Affluent Washington suburbs now face an inundation of out-of-state garbage trucks on their way to feed the "deepening dependence of central Virginia counties on revenue from landfills."[63] In Virginia's Charles City County, for example, some $32 million in revenue has already been gained from a 934-acre landfill operated by Waste Management, which has been handling increasing quantities of New York City's waste. The disposal of New York's waste has contributed to the funding of the local school system and lowered property taxes in one of the poorest counties in Virginia where almost two-thirds of the population is black. Issues of race, class, and poverty intersect in rural Virginia in a very different way from the highly policitized context of New York City, yet the disposal of waste in poor rural black districts was one of the key issues raised by the environmental justice movement when it first gained national prominence in the 1980s. The newly emerging geography of waste management now operates at a far wider scale, beyond the kind of specific locational disputes that have preoccupied neighborhood-based urban environmental groups. With the increasing significance of political campaign money from the powerful waste management industry, it appears that the regional and national dynamics to the waste economy will be far less amenable to local political control.

If New York City comes to rely on this long-distance solution, it will require the construction of a series of marine transfer stations in Greenpoint-Williamsburg, Sunset Park, Red Hook, and Hunts Point, ushering in a new phase in the city's fractured waste politics as the city signs twenty-year contracts with giant private-sector waste handlers such as Waste Management and BFI/Allied Waste.[64] New community-based waste groups have emerged such as Neighbors Against Garbage, Boroughs Allied for Recycling and Garbage Equity, and the Organization for Waterfront Neighborhoods, which have explicitly focused their campaigns against powerful waste management corporations that have been subjected to a plethora of court actions across the country because of their poor environmental record.[65] In the South Bronx, for example, there has been local outcry against a proposed waste transfer facility to be operated by the much-maligned Waste Management in a neighborhood that already has two dozen waste transfer stations, while in Williamsburg local activists have likened the stench from garbage trucks to "a million elephants" as rates of haulage have sharply increased under the city's short-term deals with an array of private waste carters to cover the closure of the Fresh Kills landfill.[66]

The defeat of urban incineration in the 1990s has opened up a new and more complex phase in the politics of pollution which has yet to be resolved. The city's post-Fresh Kills waste strategy has involved widening the involvement of major private waste companies in order to shift operational responsibilities out of the public sector (a parallel development to the mooted water privatization discussed in chapter 1) but also to undermine the power of organized crime in the vastly expensive and inefficient arena of Mob-controlled commercial waste disposal. The culture of the waste management industry has been changing from a dominance of family-run small firms with local or regional monopolies toward a new generation of powerful international corporations that can offer greater economies of scale and new sources of technical expertise. Thus the environmentalist emphasis on recycling and municipal household waste captures only a partial dimension to the changing waste scene: the technical, economic, and political barriers to recycling and waste reduction mean that landfill will play a major long-term role in the post-incineration era. The growing power of neighborhood-based waste politics forms one facet of a new political economy of urban waste in which profit margins, property values, and environmental law have combined to redistribute waste externalities onto poorer communities elsewhere.

5.3 Reclaiming the Social Environment

> We, the Asian, Native, Latino, African American and low income communities of New York City, come together to determine our environmental issues, needs and goals toward seeking justice. . . . We recognize that the fate of all people depends upon our ability to end all forms of racism and inequality and to preserve and protect the life giving forces of the earth for future generations.
>
> —*The New York City Environmental Justice Alliance*[67]

In the 1960s, wide-ranging coalitions of labor, consumer, religious, and environmental organizations emerged to rival the successful public health movements of the past. By the early 1970s the American environmental movement had become sufficiently powerful to force the passage of a number of new laws such as the 1970 Clean Air Act, the 1972 Clean Water Act, and the 1970 Occupational

Safety and Health Act.[68] In New York this phase of legislative change led to significant environmental improvements in fields such as water pollution, lead emissions, and the creation of wildlife sanctuaries.[69] Despite these advances, however, the reformist legislative agenda of the late 1960s and early 1970s never directly tackled the root causes of environmental degradation and health inequalities in American society. By the 1970s important tensions emerged between the socioeconomic imperatives of the labor movement and the increasingly antiproductivist stance of the mainstream environmental movement. The combination of economic recession with neo-Malthusian demands for zero growth precluded any kind of enduring political alliance between labor and the burgeoning environmental movement. The latter now increasingly reflected the recreational demands of a growing middle class focused on quality of life issues rather than the concerns of economic security fundamental to the New Deal era. The combination of deindustrialization with more populist neoliberal forms of political activism began to shatter the coherence of the environmental coalitions of the past. "As the movements of the sixties declined," writes Mike Davis, "new movements of the right surged forward":

> Black power and women's liberation were eclipsed by middle-class militancy, as unprecedented numbers of white-collar, professional and managerial strata became active in single-issue campaigns or local politics, often abandoning their old party affiliations en route. . . . This "greening" of American politics effectively disenfranchised the poor, while simultaneously ensuring that the new activism of the middle classes acted as a ventriloquism for the voices of corporate PACs [political action committees] and New Right lobbies.[70]

During the 1980s a further series of tensions emerged in American environmentalism. The postwar regulatory framework, exemplified by the creation of the federal Environmental Protection Agency, had operated under the implicit assumption that the impact of environmental legislation was geographically and socially neutral. This perception was shattered by a series of studies revealing how facilities such as hazardous waste landfills and noxious industrial plants were disproportionately concentrated in poor and minority-dominated communities. The 1980s saw a series of high-profile political campaigns waged by minority communities against noxious industries in their neighborhoods, which led in 1991 to the First National People-of-Color Environmental Leadership Summit in Washington, D.C.[71] A whole series of new alliances emerged, both geographically and culturally, bringing together disparate communities in order to challenge the health-threatening effects of poor environments. The environmental justice movement can be differentiated from the mainstream environmental movement in that its many hundreds of local and regional organizations are typically led by "women, working-class people, and people of color," who have been drawn mainly from America's urban ghettos, barrios, Native lands, and poor rural enclaves such as Appalachia.[72] In contrast, the established environmental movement is overwhelmingly white and middle-class and has devoted much of its energies to legislative change and the preservation of nature. There is a tension, therefore, between the political radicalism of the environmental justice movement and the established environmental movement's concern with "institutional consensus, compromise, and professionalization."[73] There is also conflict over the meaning of American landscape and culture emanating from the racist origins of the conservation movement and its enduring ambivalence toward modernity, urbanism, and cultural diversity.[74] The protection and restoration of "authentic" American landscapes has tended to obscure alternative conceptions of nature and landscape design. A reactionary pastoral vision has infused the dynamics of urban decline and racial segregation to produce new kinds of bourgeois utopias that use bucolic motifs to disguise their own artificiality. The regional organicism of twentieth-century American environmentalism has evolved in an uneasy relation to "nativist" doctrines that can all too easily be translated into a fear of strangers, whether plants or people.[75] Jens Jensen, for example, though a leading modernist in the movement for more progressive planning ideals in interwar America, drew on a racialized topography of environmental differences to bolster the perceived cultural superiority of European attitudes toward nature.[76]

It is in cities, however, that the tensions between ecological doctrine and social justice have become most apparent. In 1967 the president of the Conservation Foundation, Sydney Howe, lamented that "we are today a racially segregated profession. . . . Conservation must be of and for increasingly urban environments and their people."[77] With the emergence of "deep ecology" and new forms of transcendental nature worship in the 1970s, an ideological divide opened up between the nineteenth-century romantic roots of postwar environmentalism and a renewed emphasis on the social construction of nature and the urban environment. William Cronon has criticized the ideological draw of wilderness as "the locus for an epic struggle between malign civilization and benign nature, compared with which all other social, political, and moral concerns seem trivial."[78] The definition and ideological resonance of nature is at stake here: is nature to be defended as an imaginary lineage from the past, or to be transformed in order to create a more progressive social order? Recent disputes within the Sierra Club over the impact of immigration on the American environment are testament to deeply reactionary dimensions to the American environmental tradition which draw on nature as an ideological resource devoid of any sensitivity to social and historical context.[79] The persistence of neo-Malthusian preoccupations with overpopulation reveal how the very terms "nature" and "environment" evoke sharply different responses from different wings of the environmental movement.

The national emergence of the environmental justice movement has had a crucial impact on the development of environmental politics in New York City. The first National People-of-Color Environmental Leadership Summit of 1991 drew many New York activists together for the first time and led to the creation of the New York City Environmental Justice Alliance. This Harlem-based umbrella organization brought together over a dozen different minority-led organizations including the South Bronx Clean Air Coalition, El Puente, the Magnolia Tree Earth Center, and West Harlem Environmental Action.[80] For Vernice Miller, cofounder of West Harlem Environmental Action, the environmental justice movement presented a challenge to historical shortcomings of mainstream environmental organizations such as the Sierra Club and the Natural Resources

Defense Council "to get off the stick of preserving birds and trees and seals and things like that and talk about what's affecting real people."[81] Like the agenda of the Young Lords of the late 1960s, the political agenda of the New York City Environmental Justice Alliance covers a range of community concerns including lead poisoning, asthma, asbestos, vehicle exhaust fumes, waste incineration, and access to public space. Yet there remain important differences between the political activism of the 1960s and the 1990s. The Young Lords were an example of an urban social movement committed to far-reaching social change, but this is hardly the case with the contemporary environmental justice movement in New York City. Randy Shaw, for example, describes the US environmental justice movement as an "activist coalition" that "combines two or more organizations in pursuit of at least one mutual objective."[82] Unlike the urban social movements of the 1960s, the contemporary environmental justice movement is devoted to a series of specific objectives rather than presenting a structural challenge to American society.

A further difference from the classic model of an urban social movement is the high degree of interaction between municipal government and new environmental initiatives within the city. As early as 1989, for example, Mayor Dinkins instituted changes to the City Charter to adopt a "fair share" criterion in the location of public facilities, making New York the first major city to adopt procedures of this kind (recognized with an award at the 1992 National Conference of the American Planning Association).[83] In a second policy innovation, grassroots environmental activists have been successful in pushing the city and state governments to set up "environmental benefits programs" to enable wider public participation in environmental policy making. The first of these environmental benefits programs was created in Greenpoint-Williamsburg, with an open membership structure allowing anyone who had attended three meetings to join. The early negotiations exposed a highly complex political agenda crisscrossed by an array of ethnic and ideological tensions. These discussions revealed a lack of consensus over whether manufacturing activity should be allowed in the local area, suggesting real difficulty in extending the environmental agenda to encompass social and economic issues such as employment creation and industrial regeneration.

Differences in priorities also emerged between the Greenpoint activists, whose concerns focused on the poorly operated Newton Creek sewage treatment plant, and those of Williamsburg whose emphasis has been overwhelmingly on the proposed Brooklyn Navy Yard incineration plant.[84]

Despite these political and organizational difficulties, the Greenpoint-Williamsburg Environmental Benefits Program has received a series of major national and international awards. A first element in this program is a two-part epidemiological study conducted by the New York City Department of Health on mortality and morbidity in the Greenpoint-Williamsburg area which explored the incidence of four health issues identified to be of greatest concern to the local community: cancer, asthma, birth defects, and childhood lead poisoning. The initial outcome of these studies, however, presented no clear evidence for higher rates of environmentally induced ill health in comparison with the rest of the city, except for localized concentrations of certain types of leukemia, stomach cancer, and asthma.[85] The science of environmental epidemiology raises a host of technical complexities that may undermine the political salience of a broader appeal to improve environmental quality and public health. Issues include the tracking of multiple pathways of exposure to chemicals, the complex synergistic effects of different substances in combination, the differential impacts on individuals, and the limited knowledge of the effects of many hundreds of chemicals on human health.[86] The complex interface between the science of epidemiology and community activism can be demonstrated by recent research into the causes of asthma that has downplayed the effects of air pollution in relation to pest allergies and other features of the domestic environment. The so-called "cockroach debate" has exposed long-standing political tensions over the stigmatization of poverty and insanitary housing (see chapter 4).[87] Research into health inequalities has also identified structural features in the provision of health care that may interact with local environmental factors.[88]

Another dimension to the Greenpoint-Williamsburg Environmental Benefits Program has been an integrated multimedia pollution prevention project to build a complete picture of air, water, and solid waste emissions from different sources. Initiatives such as the Toxic Release Inventory, a federal requirement since 1987 in the wake of the Union Carbide explosion at Bhopal, India, have enabled the grassroots monitoring of sources of pollution to an unprecedented extent.[89] This emphasis on "lay science" or "popular epidemiology" has emerged as a critical dimension to new forms of environmental regulation that seek to improve the legitimacy, scope, and cost-effectiveness of environmental regulation. "Scientific knowledge acquired through actual participation," notes the academic and environmental activist Michael K. Heiman, "becomes a part of a people's culture, no longer an alien product to be accepted as an article of faith."[90] Further elements of this environmental benefits program include the development of a sophisticated geographical information system (GIS) in collaboration with university research teams to provide a detailed display of environmental, land use, health, and financial data at a lot-by-lot scale; the appointment of an urban environmental "watchperson" acting as an independent educator and advocate; and the funding of an array of environmental education projects, including innovative work at the Latino educational foundation El Puente that has channeled youth activism toward social and environmental issues.[91] In effect, these new initiatives have contributed toward the development of an environmentally informed public sphere that has further eroded the model of technical expertise inherited from the Progressive era of American environmentalism.

The Greenpoint-Williamsburg model of community mobilization has set a significant precedent for other local struggles in the city: the Harlem fight against the North River sewer plant and the South Bronx conflicts over sewage treatment works, medical waste incineration, and paper production all owe much of their organizational and political skill to the successful defeat of waste incineration in Brooklyn.[92] This new level of political sophistication by poor neighborhoods across the city marks a shift in the politics of pollution. With the colonization of formerly undesirable areas through gentrification pressures, we find that neighborhood change becomes locked in through rising property values, leading to intensifying environmental inequalities at a regional scale. The 1990s has seen a range of new institutional structures, legislative developments, and technological applications, but these new initiatives may be vulnerable to future political and economic shifts in a similar fashion to the government-funded programs of the

5.6 Mural in Bushwick put up by the El Puente youth group, circa 1993.
Source: Courtesy of Suzanne Tobias.

1960s and 1970s. The funding of environmental benefits programs, for example, has depended on permit violations at sewage treatment plants, which is an unusual basis for ensuring long-term continuity in funding.[93] We can argue that these programs are "transformative" to the extent that they raise community awareness and improve the responsiveness of regulatory agencies. Yet the advent of environmental benefit programs and other similar initiatives may be approaching the limit of what can be achieved without more fundamental change in American society. The increasing use of geographical information systems and more sophisticated approaches for the monitoring of spatial inequalities may serve to demystify the practice of environmental regulation, but greater access to information does not in itself alter the balance of social power. The local implementation of federal initiatives such as the Community Right-to-Know Act and the Toxic Release Inventory have enabled new sources of data to enter the public domain, but the long-term political and geographical consequences for the location of hazardous facilities remains uncertain. The extension of New York's ecological frontier to the landfills of Pennsylvania or Virginia suggests that the impetus of the environmental justice movement in the absence of structural changes in the organization of production may lead to widening pollution disparities across America.[94] Much has been made in recent years of the expanded use of geographical information systems, the Internet, and other technological tools for the purposes of political empowerment, but the fundamental dynamic behind the production of ecological and social devastation for marginal communities remains largely unchanged.[95]

5.4 Trash Can Utopias

The technical and political complexity of the post-Fordist waste economy is matched only by the ideological obfuscation of its alternatives. The uncertainty surrounding the future of New York's waste stream after the closure of the giant Fresh Kills landfill indicates how a perceived waste crisis can play a powerful role in driving through changes in public policy. The defeat of the Brooklyn incineration plant suggests that the politics of urban waste has entered a new and uncertain phase. The environmental contradictions of the post-Fordist waste economy can no longer be resolved within New York City but must now entail complex negotiations with other places over ever greater distances. The success of the anti-incineration protest in New York City has not fundamentally altered the dynamics of waste production but has simply shifted its spatial consequences along a changing axis of social power.

Even recycling, the discredited panacea for urban waste management, raises its own specters of sociospatial equity: at a city level there has been a bitter dispute over a proposed paper recycling plant in the Bronx, and at an international level there is evidence of poor working conditions at recycling plants for batteries, plastics, and other materials in Mexico, South Asia, and elsewhere.[96] Evidence suggests that the New York recycling program has cost the city an additional $500 million since its inception in the early 1990s, to say nothing of extra traffic movements, the congestion of building space, and wasted hours spent sorting materials that have little or no economic value.[97] At the city's ten recycling plants, teams of low-wage, mostly immigrant workers pick through mixed household waste in conditions little better than that endured by the ragpickers and scow trimmers of the nineteenth century. The Browning-Ferris-owned Bushwick plant in Brooklyn is typical, with its 195 workers, mainly recent immigrants from Central America, Africa, and the Dominican Republic, working long hours to sort through the detritus of the metropolis. It seems that New Yorkers are not very adept at distinguishing between different kinds of waste, as a third of the materials must be discarded rather than recovered: "A result is the dirty, motley mess of stuff, with a high percentage of unusable and sometimes undefinable items that the Browning-Ferris workers must sort each day, a job that most describe as tiring, repetitive and boring."[98] The hand-sorting of waste in the twenty-first century is more reminiscent of the gritty nineteenth-century city of Jacob Riis than a glimpse into a high-tech, sustainable urban future. The structural dimensions to the post-Fordist waste economy remain unchanged under the ideological cloak of postconsumer recycling. The advocacy of recycling as an end in itself involves no critical engagement with the processes of waste production, which are perpetuated under the end-of-pipe discourses of green consumerism and ecological modernization.[99] The progressive impulse of environmental reform has become preoccupied with a narrow range of targets and indicators rather than the articulation of a wider vision for urban society.

The example of Greenpoint-Williamsburg illustrates a new kind of relationship between different tiers of municipal governance and civil society. The failed imposition of a secretive, expert-led, and highly centralized policy agenda has led to an opening out of regulatory and policy-making processes. This changing emphasis has in part been driven by the attempt to gain greater public acceptance and understanding of the complexities of environmental policy making, but it has also been necessitated by the perceived exhaustion of established regulatory procedures. A recent exchange of views in the *Boston Review* illustrates some of the dilemmas posed by new kinds of environmental regulation. In an article focusing on the example of Chesapeake Bay, law professors Charles Sabel, Archon Fung, and Bradley Karkkainen stress how "the new framework forces continuous improvements in both regulatory rules and environmental performance while heightening the accountability of the actors to each other and the larger public."[100] They argue that "it is impossible to obtain the panoptic knowledge required by the Madisonian ideal" because the scale of environmental systems generates so many uncertainties. In place of centrally directed regulatory intervention they advocate locally directed forms of environmental regulation in order to alter the relationship between technical experts and the public in the interests of efficiency, flexibility, cost-effectiveness, and democratic participation.[101] The proposed regulatory model seeks to address scientific uncertainty through extended patterns of public participation, in a radical realignment of the scope and pretext for environmental policy making. In a sense these arguments bring us full circle to the critical issues raised by the emergence of the antitoxics movement, the campaign for environmental justice, and the earlier failure of the Brooklyn incinerator project's Citizens' Advisory Committee of the early 1980s. The dilemma is that the devolution of policy deliberation does not involve a fundamental shift in the balance of social power: particular places may be relatively empowered within a changing patchwork of community-based mobilization, but the overall dynamics of environmental degradation remain in place. The decentralization of the policy process forms part of a broader dissolution of existing patterns of regulation that has actually made it more difficult to articulate a public interest that conflicts with the existing parameters of a reformist and incremental political agenda. The environmental activists Matt Wilson and Eric Weltman, for example, suggest in their reply to the arguments of Charles Sabel and his colleagues that a reliance on "engaged citizenship" may risk an abdication of government responsibilities as part of a broader deregulative momentum that cannot favor environmental equity.[102]

Despite the limitations of Sabel, Fung, and Karkkainen's argument, this exchange of views illuminates a series of developments that will play a significant role in shaping emerging patterns of environmental regulation. The limits to both the centrally directed Hobbesian model and its neoliberal antithesis necessitated the emergence of new kinds of interaction between government and civil society in the interests of policy legitimacy and regulatory effectiveness. The ongoing redefinition of relations between experts and civil society rooted in the critique of urban planning since the 1960s has also opened up aspects of policy making to much greater public scrutiny than in the past. Still unresolved are how the fiscal and political parameters of the environmental debate are generated in the first place and the extent to which the rhetoric of "engaged citizenship" belies a wider deregulative momentum that is inimical to the public interest. To what extent, for example, do these new regulatory mechanisms represent a procedural disengagement from more substantive ethical issues? It is this question of ethics that underlies the significance of the environmental justice movement and the uneasy tension with new patterns of public participation.

The environmental justice movement is thus engaged in a delicate balancing act between the continuing dynamism of its grass roots and the limited interventions offered by newly emerging institutional mechanisms for environmental regulation. Apart from the immediate focus on public health and cooperation over aspects of social and environmental policy, there is little sense in New York of a more broadly articulated challenge to dominant social institutions. The

Brooklyn anti-incineration campaign shares more similarities with the classic Alinsky model of neighborhood protest than with the political radicalism of the Young Lords or other examples of urban social movements drawn from urban history. Yet the dynamics of local environmental protest involved a high degree of organizational innovation in order to build an activist coalition across the ethnic divisions of Greenpoint-Williamsburg. While highly significant in terms of community relations, this coalition does not form part of a wider movement for social change. The Brooklyn anti-incineration campaign trounced a whole gamut of political and economic elites in the city, but this victory did not serve as an opening for an alternative political program capable of rivaling existing party machines. The environmental justice movement as it has manifested itself in New York lacks the ideological coherence to present an alternative urban vision except on a small scale through the transformative work of organizations such as the grassroots educational academy El Puente or the Bedford-Stuyvesant-based Magnolia Tree Earth Center.[103] There is as yet no citywide (or even boroughwide) movement capable of challenging existing political and social structures: the gulf between grassroots radicalism and the institutionalized atrophy of electoral politics remains immense.

The environmental justice movement has identified a series of connections between class, gender, race, and age, which structure social engagement with the urban environment as a powerful alternative to mainstream environmentalism. Yet the conceptualization of the connections between racism and environmentalism remain in their infancy: misleading dualisms persist in evaluating the relative roles of race and class in determining degrees of exposure to environmental pollution; and the political ambiguities of populist antiglobalization sentiments remain only partially explored.[104] Even if environmental justice struggles have succeeded in highlighting the structural dimensions to pollution and environmental inequality, these connections have often been obscured or overlooked in the largely empirical academic literature. The environmental justice movement's focus on places of residence rather than places of work underlies the disjuncture between production and consumption that dominates environmental politics: the main difference between the mainstream environmental movement and its detractors is rooted in the tension between the protection of poor neighborhoods and the protection of affluent neighborhoods.

Debates surrounding environmental justice have highlighted tensions between different conceptions of social justice in urban space: the long-standing antinomy between liberal and Marxist interpretations (exemplified by the work of John Rawls and David Harvey respectively) has left much of the institutional constitution of the modern city only partially understood.[105] The liberal tradition has, for example, neglected the precise material and historical contexts within which particular conceptions of justice and rights have emerged. Yet the emergence of political liberalism as a form of "political and civic pluralization" is integrally related to the history of urbanization and the long-standing role of cities as centers of freedom from persecution.[106] For Ira Katznelson, the positivist dominance of the liberal tradition in urban studies has obscured the wider lineage between political liberalism, urbanism, and the emancipatory promise of modernity. Furthermore, the liberal-Marxist divide has too easily been translated into a contrast between pattern and process in urban studies, thereby neglecting other potential sources of insight. Both liberal and Marxist traditions in urban theory have struggled to bridge the space between theoretical abstraction and the lived space of the city.[107] Yet the political salience of urban social struggles derives from the reformulation of ideas in concrete situations rather than the refinement of rules of social conduct or the verification of an inner logic of social transformation.

At a conceptual level this chapter has begun to dissect some of the ways in which poverty and race intersect in the commodifaction of space and the production of the social environment. Much of the political conflict we have examined has been rooted in community-based forms of grassroots political action. Repeated reference has been made to the term "community," but this is by no means a simple or self-evident unit of social action: a survey of the sociological literature in the early 1950s, for example, uncovered more than ninety meanings of the term.[108] Despite these semantic uncertainties, however, the term has a peculiar resonance for urban environmental conflict whether in relation to urban planning, air quality, or any other dimension of dwelling-focused political mobilization. As the philosopher Iris Marion Young points out, not only does an emphasis on the community privilege small-scale, decentralized, face-to-face interactions but "it often operates to exclude or oppress those experienced as different."[109] We saw how the strength of the Brooklyn antiincineration movement was founded on the transcending of community as an exclusionary social form rooted in a shared ethnicity: the sense of place-bound identity was to predominate over a more narrowly group-based form of communal identity. The challenge for grassroots political action is to transcend both the parochialism of community and the liberal individualism of the market in order to articulate an inclusive agenda based on human diversity, creativity, and interaction.[110] In this sense ecological enlightenment and democratic urbanism are linked, because exclusionary conceptions of environmental justice merely transfer pollution onto less powerful communities.

Epilogue

On a July evening in the summer of 2000 the gates of Central Park were closed. Helicopters whirred overhead and many people stayed indoors. A little-known but deadly pathogen, the West Nile virus, had produced a modern equivalent of the yellow fever scares of the early nineteenth century. The lakes and water bodies of the park had become breeding grounds for the blood-sucking mosquitoes that pass the virus on to birds, other animals, and humans. Public debate oscillated between demands for an intensified pesticide spraying program and fears over the carcinogenic properties of the pesticide malathion for human health. An intricately designed naturalistic setting, complete with shady bosks and winding woodland paths, faced the unexpected intrusion of wild nature. The example of the West Nile virus serves as a poignant reminder that our distinctions between nature and culture, and between "good" and "bad" forms of nature, are the fluid outcome of different social and historical processes. New threats of disease remind us that the biological protagonists of urban history have not disappeared amid the virtual spaces, shopping malls, and architectural simulacra of the contemporary city.

In our survey of New York City we have encountered different cultural appropriations of nature ranging from early bacteriological advances to recent organizational innovations in pollution control. We have explored successive attempts to alter the role of nature within the city and the different ways in which conceptions of nature reflect wider political and ideological discourses. Urban nature has emerged as a hybrid entity produced through processes of social and cultural transformation. We can attempt to impose sharp analytical distinctions or clearly differentiated historical periodizations, but our object of study will always succeed in eluding attempts to subsume the production of urban space within the confines of any narrowly conceived conceptual framework. The very complexity of the terms "nature," "landscape," and "environment" are testament to the rich diversity of social and cultural traditions bound up with the histories of capitalist urbanization.

Since the origins of the first cities, those who built and planned them have often aspired to emulate perfection in nature as a blueprint for urban design. The aqueducts and gardens of ancient Rome forged a powerful dialectic between urban form and civil society that has persisted in Western design traditions into the modern era. The fountains of Renaissance Italy and the squares and gardens of Georgian London all used nature to embellish urban design and attest to the taste and civic virtue of wealthy patrons. This was nature transformed in the service of cultural ideals far removed from the demands of material subsistence. And beyond the Western tradition we can find countless other examples of intricate interactions between nature and cities, such as the water-gathering technologies of the Islamic world or the elemental geometries of early Chinese urbanism. The interpretation of these landscapes tells us much about the structures and beliefs of the societies and their selective appropriation of the material and symbolic wealth of nature.

The modification of nature and landscape enters a new phase in the modern era. The creation of metropolitan nature marks the emergence of a new synthesis between technology and urban design. In nineteenth-century Europe and North America we find a celebration of the utilitarian aesthetic of public health engineering in response to the challenges of capitalist urbanization. The aqueducts and water tunnels of Jervis and Chesbrough, the sewer systems of Bazalgette and Belgrand, and the parks of Olmsted and Alphand all form part of an unfolding dynamic of modern urbanism. In place of the chaotic and dangerous nineteenth-century city, a new kind of engagement emerged between controlled nature and the modern citizen. Yet the refinement of urban planning ideals in the nineteenth century reveals an ambiguity between the notion of some kind of natural order or symmetry achievable in urban form and the competing discourses of a "naturalized" economic order rooted in laissez-faire political economy. The power of professional expertise to impose particular conceptions of urban design stems from the unique conjunctions of political power, economic prosperity, and legal innovation that emerged in the nineteenth-century industrial city. In the twentieth century the advent of scientific urbanism and technological modernism heralded ever more ambitious attempts to mold urban space in the service of new social and political ideals which sought to dispense with the spatial disorder of the past. The development of distinctive cultures of taste imbued technical discourses with a sense of intellectual superiority over rival perspectives on the form and function of modern cities.

The period from the middle decades of the nineteenth century until the 1970s marks a phase of relative stability in relations between nature, capital, and urban space. Successive economic cycles strengthened rather than weakened the organizational and political pretext for the reconstruction of urban infrastructure. With the growth of new forms of civic consciousness since the nineteenth century and the gradual widening of the political franchise, reworked elements of nature became part of a democratic urbanism in the service of a far wider public than the private and ornamental landscapes of the past. The human body and engineering science were combined to produce a cyborg urbanization surpassing earlier forms of urbanism in its complexity and ingenuity. Nature played a pivotal role in the development of American modernism, as a means to articulate a more inclusive sense of national identity through the production of public landscapes at the leading edge of engineering technology and landscape design. The "democratic pyramids" of the New Deal era, exemplified by state-of-the-art water technologies and ambitious regional planning idioms, were the ultimate expression of a modernist ethic that sought to produce a new kind of progressive mediation between society and nature.

For critics of modern urbanism, however, the extending ecological frontier was conceived in conspiratorial terms as a variant on the despotic hydraulic societies of the past. The richness of the modernist legacy in landscape design was overwhelmed by the technocratic advance of an urbanism that had lost sight of any clearly defined progressive agenda. In the post-World War II era the design legacies of modernist pioneers such as Garrett Eckbo and Richard Neutra became subsumed within the new urban landscape of corporate America. The expert vision inherited from the Progressive era struggled to maintain its legitimacy from the 1960s and 1970s onward: attempts to foster greater public participation in decision making seemed only to heighten conflict and uncertainty; and the deployment of ever more elaborate methodologies to provide a veneer of scientific objectivity to urban governance ignored the political dynamics of the civil rights era. The gradual realization that the imposition of any centrally directed urban vision conflicted with the heterogeneous realities of urban life ushered in more diffuse patterns of municipal governance.

The ultimate collapse of the Lindsay era of New York politics is testament to this contradictory dynamic between greater social inclusiveness and the underlying impetus of capitalist urbanization. By the 1970s New York City faced a plethora of challenges emerging from a tension between its dilapidated physical fabric and its expanded global economic role. The city seemed increasingly comprised of irreconcilable elements; a maelstrom of competing demands left existing conceptions of urban and regional planning in disarray. The fiscal crisis of the 1970s tore at the heart of the public city, producing ever-widening social and spatial disparities. Yet this process was not driven but merely facilitated by processes of globalization, as regional political and economic elites sought to dismantle the public realm of the New Deal era. The recapitalization of New York City under Koch was simply the outward manifestation of a brutal realignment of city politics to serve the needs of the emerging FIRE economy and create a new business environment more conducive to the needs of international capital. Writing in the early twentieth century, Henry James reflected on New York's "perpetual repudiation of the past," a sentiment that reached its acme with the erasure of whole communities under the vast infrastructure projects of the Moses era. It seems, however, that the contemporary city is intent on quite the opposite path, whether through the aesthetic emulation of the past or the recreation of the vast disparities in wealth that marked the early twentieth-century metropolis. The physical transformations of the past have been replaced with new forms of social engineering advanced through the combined effects of gentrification and the marketability of urban chic. The triumphant return of the private city now threatens to undermine any meaningful basis for an invigorated public realm and its physical expression in urban space.

The presence of nature in cities has served to naturalize the urban process in a variety of ways. Indeed nature has been a vital part of the dynamics of capitalist urbanization from the outset. We saw how Central Park was never an anticapitalist oasis but from its inception a sophisticated dimension to the commodification of nature. Similarly, the capture of the regional hydrological cycle through the construction of an elaborate water supply system underpinned the expansion of the city and set in place new configurations for the administration and funding of urban infrastructure. The very idea of urban nature is a contested terrain that intersects with the dynamics of the urban process at every level. The sharpest divide has been between the idea of nature as an invariant essence residing outside of social relations and the idea of nature as socially and historically constituted. This tension translates into a distinction between conceptions of urban form as the outcome of the single-minded genius of individuals and an emphasis on the social production of space through the collective imprint of society. As cities have lost their former functions as sites of production, this has profoundly altered the meaning of urban landscapes. The changing social complexion of postindustrial places has reconstructed the spatial dynamics of the politics of nature. The regional organicism and "nativist" landscapes associated with the emergence of American modernism now provide little resonance for a more inclusive urban design ethic that is not easily reducible to the cultural heritage of any particular place or region. The emerging Latino metropolis, for example, imbues North American cities with a very different set of meanings from the iconographic legacies of the past. The earlier strands of German romanticism, the English picturesque, and Italian futurism are now overlaid with the vernacular urbanism of Latin America, the rediscovery of Antillean landscapes, and a myriad of other influences.

The contemporary juxtaposition of cities and nature has produced some powerful incongruities. It is as if the premodern, the modern, and the postmodern are all jostling for position simultaneously: environmentalists call for a return to cyclical patterns of resource use as part of a premodern bioregional ethic; the vast urban infrastructures built up over the modern era continue to provide material sustenance for urban society in spite of their neglect; and historic cultures of nature serve as aesthetic enhancement for the postmodern simulacra of urban real estate. Urban nature is a collage of past and present, a medley of different elements that binds the concrete fabric of the city to the abstract commodification of space. In New York City we are confronted with successive layers of urban change: abandoned lots have become community gardens brimming with flowers and vegetables from all over the world; the sparks from the subway illuminate a kaleidoscope of graffiti art as public spaces are continually created and destroyed; and mountain water makes its way to the city through a complex of pipes and aqueducts constructed over many decades. The clouds of steam rising from the street remind us that the possibilities for urban life are sustained by an unseen web of structures, connections, and relationships.

Notes

Introduction

1

Lewis Mumford, *The city in history: its origins, its transformations, and its prospects* (London: Secker and Warburg, 1961), p. 571.

2

John Dos Passos, *Manhattan transfer* (Leipzig: Bernhard Tauchnitz, 1932), p. 15.

3

Cited in "Voyage of John de Verrazzano along the coast of North America from Carolina to Newfoundland (containing the first discovery of Hudson's river)," trans. Joseph G. Cogswell, in G. M. Asher, *Henry Hudson the navigator* (London: Hakluyt Society, 1860), p. 203; and Isaac Newton Phelps Stokes, *The iconography of Manhattan Island, 1498–1909,* vol. 1 (New York: Robert H. Dodd, 1915), p. 3.

4

This argument is powerfully developed with respect to nineteenth-century France by the art historian Nicholas Green, who shows that nature "was an historically specific construct, a product of discourses materially grounded in the conditions of contemporary Paris." Nicholas Green, *The spectacle of nature: landscape and bourgeois culture in nineteenth-century France* (Manchester: Manchester University Press, 1990), p. 184.

5

See Rosalyn Deutsche, "Property values: Hans Haacke, real estate, and the museum," in *Evictions: art and spatial politics* (Cambridge, Mass.: MIT Press, 1996), pp. 159–192.

6

Karl Marx observed in his essay "The fetishism of the commodity and the secret thereof": "It is as clear as noon-day, that man, by his industry, changes the forms of the materials furnished by Nature, in such a way as to make them useful to him." Karl Marx, *Capital,* vol. 1, trans. Samuel Moore and Edward Aveling (1867; London: Lawrence & Wishart, 1974), p. 76.

7

For an example of a narrowly conceived conception of landscape rooted in "the traditional love of agriculture and distrust of artifice," see John R. Stilgoe, *Common landscape of America, 1580 to 1845* (New Haven: Yale University Press, 1982), p. 345.

8

We should note that the idea of ideology is laced with ambiguity; it may mean anything from a concern with "ideas of true and false cognition" to a more dynamic emphasis on the general "function of ideas within social life." See Terry Eagleton, *Ideology: an introduction* (London: Verso, 1991), p. 3. We are perhaps better served by John B. Thompson's understanding of ideology as the way in which "meaning (or signification) serves to sustain relations of domination" (ibid., p. 5). The centrality of nature to the meaning of ideology is stressed by Joachim Wolschke-Bulmahn who argues that "nature is ideology" in his survey of the politics of landscape design. See Joachim Wolschke-Bulmahn, ed., *Nature and ideology: natural garden design in the twentieth century* (Washington, D.C.: Dumbarton Oaks Research Library and Collection, 1997), p. 6.

9

Thomas Jefferson to Benjamin Rush, 23 September 1800, in *The works of Thomas Jefferson,* ed. Paul L. Ford (New York: G. P. Putnam's, 1904–1905), vol. 9, p. 147.

10

See, for example, Lawrence Buell, *The environmental imagination: Thoreau, nature writing, and the formation of American culture* (Cambridge, Mass.: Harvard University Press, 1995); Jane Bennett, *Thoreau's nature: ethics, politics, and the wild* (London: Sage Publications, 1994); John B. Jackson, "Jefferson, Thoreau & after," in *The selected writings of J. B. Jackson,* ed. Ervin H. Zube (Amherst: University of Massachusetts Press, 1970), pp. 1–9.

11

See Henri Lefebvre, "Town and country" (1965), in his *Writings on Cities,* ed. and trans. Eleonore Kofman and Elizabeth Lebas (Oxford and Cambridge, Mass.: Blackwell, 1996), pp. 118–121; Raymond Williams, *The country and the city* (Oxford: Oxford University Press, 1973).

12

Richard Sennett, *Flesh and stone: the body and the city in Western civilization* (London: Faber and Faber, 1994). See also Graeme Davison, "The city as a natural system: theories of urban society in early nineteenth-century Britain," in Derek Fraser and Anthony Sutcliffe, eds., *The pursuit of urban history* (London: Edward Arnold, 1983), pp. 349–370; and Caroline van Eck, *Organicism in nineteenth-century architecture: an inquiry into its theoretical and philosophical background* (Amsterdam: Architectura & Natura Press, 1994).

13

Abel Wolman, "The metabolism of cities," *Scientific American* 213 (1965): 179.

14

See Gilles Deleuze and Félix Guattari, "City-state," in Neil Leach, ed., *Rethinking architecture: a reader in cultural theory* (London: Routledge, 1997), pp. 313–316. See also Manuel De Landa, "Deleuze, diagrams, and the genesis of form," *NYA: New York Architecture* 23 (1998): 30–34; Gilles Deleuze, "The fold: Leibniz and the baroque," *Architectural Design* 19 (1993): 16–21; Gilles Deleuze, "The object of ecosophy," in Amerigo Marras, ed., *ECO-TEC: architecture of the in-between* (New York: Princeton Architectural Press, 1999), pp. 11–20; Gijs Wallis de Vries, "Deleuze en de architectuur: aanzet tot een gebruiksaanwijzing," *Archis* 11 (1993): 54–65.

15

Donna Haraway, cited in Erik Swyngedouw, "The city as a hybrid: on nature, society and cyborg urbanization," *Capitalism, Nature, Socialism* 7 (1997): 65. Haraway notes that the term "cyborg" originated in relation to the development of space travel in the early 1960s as "the enhanced man who could survive in extraterrestrial environments." Donna J. Haraway, *Modest witness: feminism and technoscience* (London: Routledge, 1997), p. 51. See also Amerigo Marras, "Hybrids, fusions, and architecture of the in-between," in Marras, ed., *ECO-TEC,* pp. 2–9; Erik Swyngedouw, "Modernity and hybridity: nature, *regeneracionismo,* and the production of the Spanish waterscape, 1890–1930," *Annals of the Association of American Geographers* 89 (1999): 443–465.

16

Alberto Abriani, "Dal sifone alla città / From the siphon to the city," *Casabella: International Architectural Review* 542/543 (1988): 117.

17

Ian McHarg, "The place of nature in the city of man," in Pierre Dansereau, ed., *Challenge for survival: land, air and water in the megalopolis* (New York: Columbia University Press, 1970), p. 54. See also Ian McHarg, *Design with nature* (Garden City, N.Y.: Doubleday, 1969).

18

Ian McHarg, "Ecology and design," in George Thompson and Frederick Steiner, eds., *Ecological design and planning* (New York: Wiley, 1997), p. 321. A critical tension between ecology and modernity has also been translated into a simplistic dichotomy between "ecological" and "mechanistic" thought exemplified by the historian Donald Worster's call to study the impact of "natural laws." On the tension between different strands in environmental history see Maureen Flanagan, "Environmental justice in the city: a theme for urban environmental history," *Environmental History* 5 (2000): 159–164; Bill Luckin, "Versions of the environmental," *Journal of Urban History* 24 (1998): 510–523; Martin Melosi, "Urban pollution: historical perspective needed," *Environmental Review* 3 (1979): 27–45; Martin Melosi, "The place of the city in environmental history," *Environmental History Review* 17 (Spring 1993): 1–23; Christine Meisner Rosen and Joel A. Tarr, "The importance of an urban perspective in environmental history," *Journal of Urban History* 20 (1994): 299–310; Jeffrey K. Stine and Joel A. Tarr, "Technology and the environment," *Technology and Culture* 39 (1998): 601–640; Donald Worster, "Transformations of the earth: toward an agroecological perspective in history," *Journal of American History* 76 (1989): 1087–1110.

19

Natasha Nicholson and Pamela Charlick, "Urban futures: city as organism," essay to accompany the exhibition "London Living City," 14 April to 8 July 2000, Royal Institute of British Architects, London. The search for natural laws in urban design is also echoed by Eliel Saarinen: "Because the principle of organic order is the underlying law of nature's architecture," wrote Saarinen, "the same principle must be recognized as the underlying law of man's architecture as well." Eliel Saarinen, *The city: its growth, its decay, its future* (Cambridge, Mass.: MIT Press, 1943), p. 18. Although other voices from within the ecological tradition of urban writing such as Jane Jacobs have been more sensitive to the daily rhythm of urban life, her neighborhood-based urbanism contained the seeds of social intolerance. See Marshall Berman, *All that is solid melts into air: the experience of modernity* (London: Verso, 1982). Indeed, the inherent political ambiguities running through Jacob's work have become even more apparent with her recent expositions on nature-emulating forms of creative capitalism as the key to urban regeneration. See Jane Jacobs,

The nature of economies (New York: Modern Library, 1999), and the review by Mike Davis, "Green streets," *Village Voice* (April-May 2000).

20

Marc-Antoine Laugier, *Observations sur l'architecture* (1765), cited in Manfredo Tafuri, "Toward a critique of architectural ideology" (1969), reprinted in K. Michael Hays, ed., *Architecture theory since 1968* (Cambridge, Mass.: MIT Press, 1998), pp. 6–35.

21

Francesco Milizia, *Principi di architettura civile* (1813), cited in Tafuri, "Toward a critique of architectural ideology," p. 11. We can also compare the biological metaphors of Milizia with the early twentieth-century writings of Patrick Geddes who sought to identify a set of universal principles for the morphology of cities. See Patrick Geddes, *Cities in evolution: an introduction to the town planning movement and the studies of civics* (London: Williams and Norgate, 1915). See also the interesting critiques of Geddes in R. Rossini Favretti, "La città 'organica': componenti cognitive e pragmatiche dell'opera di Patrick Geddes," *Parametro* 14 (1983): 88–89, 96; Volker M. Welter, "Patrick Geddes and the city as organic unity," *Edinburgh Architectural Review* 22 (1995): 11–30.

22

See Tafuri, "Toward a critique of architectural ideology."

23

Michael Sorkin, "Ciao Manhattan," in *Exquisite corpse: writings on buildings* (London: Verso, 1991), p. 358.

24

Herbert Girardet, *The Gaia atlas of cities: new directions for urban living* (London: Gaia Books, 1992). Other recent examples of "ecological urbanism" include Michael Hough, *City form and natural process* (New York: Van Nostrand Reinhold, 1984); David Nicholson-Lord, *The greening of cities* (London: Routledge and Kegan Paul, 1987); David Gordon, *Green cities: ecologically sound approaches to urban space* (Montreal: Black Rose, 1989); Victor Papanek, *The green imperative: ecology and ethics in design and architecture* (London: Thames and Hudson, 1995); Rutherford H. Platt, Rowan A. Roundtree, and Pamela C. Muick, eds., *The ecological city: preserving and restoring urban diversity* (Amherst: University of Massachusetts Press, 1994); Miguel Ruano, *Ecourbanismo: entornos humanos sostenibles* (Barcelona: Gustavo Gili, 1999); and H. Sukopp, M. Numata, and A. Huber, *Urban ecology as the basis of urban planning* (Amsterdam: SPB Academic Publishing, 1995). For an interesting critique of "bioregionalism" see Donald Alexander, "Bioregionalism: science or sensibility?," *Environmental Ethics* 12 (1990): 161–173.

25

Mathis Wackernagel and William Rees, *Our ecological footprint: reducing human impact on the earth* (Gabriola Island, B.C., and Stony Creek, Conn.: New Society Publishers, 1996), p. 7.

26

See, for example, Egon Becker and Thomas Jahn, eds., *Sustainability and the social sciences: a cross-disciplinary approach to integrating environmental considerations into theoretical reorientation* (London: Zed Books, 1999); Thomas Jahn, "Urban ecology: perspectives of social-ecological urban research," *Capitalism, Nature, Socialism* 7 (1996): 95–101; Roger Keil, "The environmental problematic in world cities," in Paul L. Knox and Peter J. Taylor, eds., *World cities in a world-system* (Cambridge: Cambridge University Press, 1995), pp. 280–297. In a similar vein, the geographer Jennifer Wolch seeks to rework our understanding of urban nature at an ontological level in order to allow the independent agency of nature to take a prominent role in our conceptualization of social-ecological relations. See Jennifer Wolch, "Zoöpolis," *Capitalism, Nature, Socialism* 7 (1996): 21–47.

27

Key examples include Murray Bookchin, *The limits of the city*, 2d ed. (1974; Montreal: Black Rose, 1986); Murray Bookchin, *The ecology of freedom: the emergence and dissolution of hierarchy* (1982; Montreal: Black Rose, 1991); Murray Bookchin, *From urbanization to cities: toward a new politics of citizenship* (London: Cassell, 1995).

28

See, for example, William Cronon, *Changes in the land: Indians, colonists, and the ecology of New England* (New York: Hill and Wang, 1983); William Cronon, "Modes of prophecy and production: placing nature in history," *Journal of American History* 76 (1989): 1121–1131; William Cronon, *Nature's metropolis: Chicago and the great West* (New York: Norton, 1991); William Cronon, "A place for stories: nature, history, and narrative," *Journal of American History* 78 (1991): 1347–1376. Other recent examples of urban environmental history that focus on the experience of one American city include Gray Brechin, *Imperial San Francisco: urban power, earthly ruin* (Berkeley: University of California Press, 1999); Greg Hise and William Deverell, *Eden by design: the 1930 Olmsted-Bartholomew plan for the Los Angeles region* (Berkeley: University of California Press, 2000); Andrew Hurley, *Environmental inequalities: race, class, and pollution in Gary, Indiana, 1945–1980* (Chapel Hill: University of North Carolina Press, 1995); and Max Page, *The creative destruction of Manhattan, 1900–1940* (Chicago: University of Chicago Press, 1999).

29

See Donna J. Haraway, *Primate visions: gender, race and nature in the world of modern science* (London: Routledge, 1989); Donna J. Haraway, *Simians, cyborgs, and women: the reinvention of nature* (London: Free Association Books, 1991), pp. 203–231.

30

Alexander Wilson, *The culture of nature: North American landscape from Disney to Exxon Valdez* (Oxford and Cambridge, Mass.: Basil Blackwell, 1992). On cultural histories of American nature see also Andrew Ross, *The Chicago gangster theory of life: nature's debt to society* (London: Verso, 1994), and David Rothenberg, *Hand's end: technology and the limits of nature* (University of California Press, 1993).

31

See, for example, Marcel Roncayolo, "Transformacions nocturnes de la ciutat. L'imperi dels llums artificials," in *Visions urbanes. Europa, 1870–1993. La ciutat de l'artista. La ciutat de l'arquitecte* (Barcelona: Centre de Cultura Contemporània de Barcelona, 1994), pp. 48–56; Sara Protasoni, "City lights: the visible and the invisible," *Rassegna* 63 (1995): 6–17.

32

See Green, *The spectacle of nature.*

33

Abriani, "Dal sifone alla città," p. 117.

34

See, for example, William J. Ronan, "Scientific management comes to New York City," *Municipal Engineers Journal* 42 (1956): 163–171.

35

Annemieke J. M. Roobeek, "The crisis in Fordism and the rise of a new technological paradigm," *Futures* 19 (1987): 129–154. See also Keller Easterling, *Organization space: landscape, houses, and highways in America* (Cambridge, Mass.: MIT Press, 1999).

36

André Guillerme, "Sottosuolo e costruzione della città / Underground and construction of the city," *Casabella: International Architectural Review* 542/543 (1988): 118.

37

Jürgen Habermas, "Modern and postmodern architecture" (1981), in *The new conservatism: cultural criticisms and the historians' debate,* trans. Shierry Weber Nicholson (Cambridge: Polity Press, 1989), pp. 3–21; Gwendolyn Wright, "Inventions and interventions: American urban design in the twentieth century," in Russell Ferguson, ed., *Urban revisions: current projects for the public realm* (Cambridge, Mass.: MIT Press; Los Angeles: Museum of Contemporary Art, 1994), pp. 26–37.

38

Jonathan Raban, *Soft city* (London: Hamish Hamilton, 1974), p. 169.

39

See, for example, Clara Weyergraf-Serra and Martha Buskirk, eds., *The destruction of* Tilted Arc*: Documents* (Cambridge, Mass.: MIT Press, 1990).

40

See, for example, M. Christine Boyer, *The city of collective memory: its historical imagery and architectural entertainments* (Cambridge, Mass.: MIT Press, 1994); Deutsche, *Evictions;* Neil Smith, *The new urban frontier: gentrification and the revanchist city* (London: Routledge, 1996); Michael Sorkin, ed., *Variations on a theme park: the New American city and the end of public space* (New York: Hill and Wang, 1992).

41

See, for example, Iain Borden, Joe Kerr, and Jane Rendell with Alicia Pivaro, eds., *The unknown city: contesting architectue and social space* (Cambridge, Mass.: MIT Press, 2001); Nan Ellin, *Postmodern urbanism* (New York: Princeton Architectural Press, 1996); Magali Sarfatti Larson, *Behind the postmodern facade: architectural change in late twentieth-century America* (Berkeley: University of California Press, 1993); and Edward W. Soja, *Postmetropolis* (Oxford: Blackwell, 1999).

42

On the interaction between the economic and cultural dimensions of postmodernity see Perry Anderson, *The origins of postmodernity* (London: Verso, 1998); Fredric Jameson, "The brick and the balloon: architecture, idealism and land speculation," *New Left Review* 228 (1998): 25–46; David Harvey, *The condition of postmodernity* (London and Cambridge, Mass.: Blackwell, 1989). On the ethical and political implications of the postmodernity debate see, for example, Richard Bernstein, *The new constellation: the ethical-political horizons of modernity/postmodernity* (Cambridge: Polity Press, 1991), and Christopher Norris, *Reclaiming truth: contribution to a critique of cultural relativism* (London: Lawrence and Wishart, 1996).

43

Kevin Lynch, *The image of the city* (Cambridge, Mass.: MIT Press, 1960), p. 2.

1

WATER, SPACE, AND POWER

1

Jean Gottmann, *Megalopolis: the urbanized northeastern seaboard of the United States* (Cambridge, Mass.: MIT Press, 1961), p. 730.

2

In addition to the official opening ceremony, the new water tunnel has also been honored by an Obie-winning multimedia art performance dedicated to the twenty-four workers known as "sandhogs" who died during its construction. See Marty Pottenger, "CWT #3: making city water tunnel #3," *High Performance* (Spring 1997), pp. 6–14.

3

Nelson M. Blake, *Water for the cities: a history of the urban water supply problem in the United States* (Syracuse, N.Y.: Syracuse University Press, 1956), p. 1.

4

Jean-Pierre Goubert, *The conquest of water: the advent of health in the industrial age,* trans. A. Wilson (1986; Cambridge: Polity Press, 1989), p. 259.

5

William Morrish, "The urban spring: formalizing the water system of Los Angeles," *Modulus: University of Virginia Architectural Review* 17 (1984): 45–83.

6

On the early history of New York City's water supply see George E. Hill and G. E. Waring, *Old wells and water-courses of the island of Manhattan* (New York: Knickerbocker Press, 1897); Irving V. A. Huie, "New York City's water supply and its future," *Municipal Engineers Journal* 37 (1951): 93–112; Gerard T. Koeppel, *Water for Gotham: a history* (Princeton, N.J.: Princeton University Press, 2000); Edward Wegmann, *The water supply of the City of New York, 1658–1895* (New York: John Wiley & Sons, 1896); Charles H. Weidner, *Water for a city: a history of New York City's problem from the beginning to the Delaware River system* (New Brunswick, N.J.: Rutgers University Press, 1974).

7

Peter Kalm, *The America of 1750: Peter Kalm's travels in North America,* ed. Adolph B. Benson (1770; New York: Wilson-Erickson, 1937), p. 133.

8

New York Journal, cited in Eugene Moehring, *Public works and the patterns of urban real estate growth in Manhattan, 1835–1894* (New York: Arno Press, 1981), p. 25.

9

Michael A. Mikkelson, "A review of the history of real estate on Manhattan Island," in Real Estate Record Association, *A history of real estate, building and architecture in New York City* (New York: Record and Guide, 1898), pp. 1–129.

10

See Alexander F. Vache, *Letters on yellow fever, cholera and quarantine; addressed to the Legislature of the State of New York* (New York: McSpedon and Baker, 1852).

11

The relatively late incidence of cholera can be related to the increasing integration of the city into transatlantic trade and travel with Europe, which also fostered the racialization of the disease in the popular imagination. It was not until Robert Koch's discovery of the bacterium *Vibrio cholerae* in 1883 that earlier moral and miasmic conceptions of the disease began to wane. The historian Charles Rosenberg writes: "Cholera, a scourge of the sinful to many Americans in 1832, had, by 1866, become the consequence of remediable faults in sanitation." Charles E. Rosenberg, *The cholera years: the United States in 1832, 1849, and 1866* (Chicago: University of Chicago Press, 1962), p. 5. See also Leona Baumgartner, "One hundred years of health: New York City, 1866–1966," *Bulletin of the New York Academy of Medicine* 45 (1969): 555–586; Gretchen A. Condran, "Changing patterns of epidemic disease in New York City," and Alan M. Kraut, "Plagues and prejudice: nativism's construction of disease in nineteenth- and twentieth-century New York City," both in David Rosner, ed., *Hives of sickness: public health and epidemics in New York City* (New Brunswick, N.J.: Rutgers University Press, 1995), pp. 27–41, 65–90. Like cholera, yellow fever was spread in the late eighteenth and early nineteenth century by ships from areas where the disease was endemic, such as the West Indies and South America. Effective public health measures succeeded in bringing the disease under control, though it continued to affect southern US ports for decades. On broader debates concerning nineteenth-century disease and urbanization see, for example, Richard J. Evans, *Death in Hamburg: society and politics in the cholera years, 1830–1910* (Oxford: Oxford University Press, 1987); Richard J. Evans, "Epidemics and revolutions: cholera in nineteenth-century Europe," in T. Ranger and P. Slack, eds., *Epidemics and ideas: essays on the historical perception of pestilence* (Cambridge: Cambridge University Press, 1992), pp. 149–173; J. N. Hays, *The burdens of disease: epidemics and human response in western history* (New Brunswick, N.J.: Rutgers University

Press, 1998); Gerry Kearns, "Biology, class and the urban penalty," in G. Kearns and W. J. Withers, eds., *Urbanising Britain: essays on class and community in the nineteenth century* (Cambridge: Cambridge University Press, 1991), pp. 12–30.

12

John W. Francis, *Letter on the cholera asphyxia, now prevailing in the city of New-York* (New York: George P. Scott, 1832), p. 29.

13

Ibid.

14

Ibid.

15

New York City Board of Health, *Report of the proceedings of the sanatory committee of the Board of Health, in relation to the Cholera, as it prevailed in New York in 1849* (New York: McSpedon and Baker, 1849), pp. 11–12.

16

The Cholera Bulletin, intro. Charles E. Rosenberg (1832; New York: Arno Press, 1972). See also Dudley Atkins, ed., *Reports of hospital physicians and other documents in relation to the epidemic cholera of 1832* (New York: Carvill, 1832).

17

Martyn Paine, *Letters on the cholera asphyxia, as it appeared in the City of New-York* (New York: Collins and Hannay, 1832). See also David Meredith Reese, *Plain and practical treatise on the epidemic cholera, as it prevailed in the city of New York, in the summer of 1832* (New York: Conner and Cooke, 1833).

18

The Common Council had been created in 1686 as the city's legislative body.

19

Huie, "New York City's water supply and its future." In 1827 there were the last recorded attempts to gain water by drilling into Manhattan Island, in yet another ill-fated private intervention under the newly formed New York Well Company. See John Duffy, *A history of public health in New York City, 1625–1866* (New York: Russell Sage Foundation, 1968), p. 391.

20

New York State, *Act of incorporation of the Manhattan Company* (New York: George Forman, 1799). The full text of the critical passage reads: "the said company shall, within ten years from the passing of this act, furnish and continue a supply of pure and wholesome water. . . . And it be further enacted, that it shall and may be lawful for the said company, to employ all such surplus capital as may belong or accrue to the said company in the purchase of public or other stock, or in any other monied transactions or operations, not inconsistent with the constitution and laws of this State or of the United States, for the sole benefit of the said company" (p. 11).

21

Duffy, *A history of public health in New York City, 1625–1866,* p. 28.

22

Ibid., p. 30.

23

New York Common Council, *Extract from the minutes of the Common Council in relation to the Manhattan Company* (New York: Childs & Devoe, 1835).

24

See Letty D. Anderson, "The diffusion of technology in the nineteenth-century American city: municipal water supply investments" (PhD dissertation, Northwestern University, 1980); Charles D. Jacobson and Joel A. Tarr, "The development of water works in the United States," *Rassegna: Themes in Architecture* 57 (1994): 37–41; Maureen Ogle, "Water supply, waste disposal, and the culture of privatism in the mid-nineteenth-century American city," *Journal of Urban History* 25 (1999): 231–347; Robert Thorne, "The hidden iceberg of architectural history," *Newsletter of the Society of Architectural Historians of Great Britain* 65 (1998): 8–9.

25

On the history of US water systems see Ellis L. Armstrong, Michael C. Robinson, and Suellen M. Hoy, eds., *History of public works in the United States, 1776–1976* (Chicago: American Public Works Association, 1976); Moses N. Baker, ed., *The manual of American water works* (New York: Engineering News, 1889); Moses N. Baker, *The quest for pure water: the history of water purification from the earliest records to the twentieth century* (Washington, D.C.: American Water Works Association, 1949); J. J. R. Croes, *Statistical tables from the history and statistics of American water works* (New York: Engineering News, 1883); Jacobson and Tarr "The development of water works in the United States"; Blake, *Water for the Cities;* J. Michael La Nier, "Historical development of municipal water systems in the United States, 1776–1976," *Journal of the American Water Works Association* (April 1976), pp. 173–180; Abel Wolman, "75 years of improvement in water supply quality," *Journal of the American Water Works Association* 48 (1956): 905–914; Abel Wolman, "Status of water resources use, control, and planning in the United States," *Journal of the Ameri-*

can Water Works Association 56 (1963): 1253–1272; Donald J. Pisani, "Beyond the hundredth meridian: nationalizing the history of water in the United States," *Environmental History* 4 (2000): 466–482.

26

Huie, "New York City's water supply and its future"; New York Committee on Fire and Water, *Report of the Board of Aldermen, Relative to Introducing into the City of New York a Supply of Pure and Wholesome Water* (28 December 1831).

27

On public health in nineteenth-century Philadelphia see, for example, Michael P. McCarthy, *Typhoid and the politics of public health in nineteenth-century Philadelphia* (Philadelphia: American Philosophical Society, 1987); Nicholas B. Wainwright, ed., *A Philadelphia perspective: the diary of Sidney George Fisher covering the years 1834–1871* (Philadelphia: Historical Society of Philadelphia, 1967); Sam Bass Warner, *The private city: Philadelphia in three periods of its growth* (Philadelphia: University of Pennsylvania Press, 1968).

28

See Donald J. Cannon, "Firefighting," in Kenneth T. Jackson, ed., *The encyclopedia of New York* (New Haven: Yale University Press, 1995), pp. 408–412. Interestingly, the last great conflagration of 1835 actually accelerated the speculative boom of the 1830s through a kind of proto-Haussmannization of the downtown business district, which could now be completely rebuilt.

29

Moehring, *Public works and the patterns of urban real estate growth in Manhattan.*

30

City bond issues were essential because the private sector was incapable of raising capital on the scale that was required. See Jacobson and Tarr, "The development of water works in the United States"; Laura Rosen, "New York builds in hard times," *Livable City* 16 (1992): 2–4.

31

Cited in Moehring, *Public works and the patterns of urban real estate growth in Manhattan,* p. 39. See also Roger Panetta, "The Croton Aqueduct and suburbanization of Westchester," in Hudson River Museum of Westchester, *The Old Croton Aqueduct: rural resources meet urban needs* (Yonkers, N.Y.: Hudson River Museum of Westchester, 1992), pp. 41–48.

32

Cited in Moehring, *Public works and the patterns of urban real estate growth in Manhattan,* p. 51.

33

Philip Hone writing in 1842, cited in I. N. Phelps Stokes, *The iconography of Manhattan Island* (1917; New York: Arno Press, 1967), p. 1777.

34

See *The diary of Philip Hone,* ed. B. Tuckerman, vol. 2, *1828–1851* (New York: Dodd, Mead & Company, 1889).

35

Vittorio Gregotti, editorial in *Rassegna: Themes in Architecture* 57 (1994): 5.

36

For six years, four thousand mostly Irish workers had worked on the largest public works project in the city's history up to that time. See Charles King, *A memoir of the construction, cost, and capacity of the Croton Aqueduct* (New York: Charles King, 1843); John B. Jervis, *Description of the Croton Aqueduct* (New York: Slamm and Guion, 1842); "The Boston miscellanies: the Croton Aqueduct," in *Being a collection of useful and entertaining articles on various subjects* (Boston: Bradbury & Guild, 1850), pp. 28–36. More recent sources include Larry D. Lankton, *The "practicable" engineer: John B. Jervis and the Old Croton Aqueduct,* Essays on Public Works History, no. 5. (Chicago: Public Works Historical Society, 1977); F. Daniel Larkin, *John B. Jervis: an American engineering pioneer* (Arnes: Iowa State University Press, 1990); George H. Rappole, "The Old Croton Aqueduct," *IA: Journal of the Society for Industrial Archaeology* 1 (1978): 15–25; Hudson River Museum of Westchester, *The Old Croton Aqueduct.*

37

Jervis's experience on the Erie Canal is significant because this was not only one of the earliest examples of large-scale public works in North America but also accelerated the trade-based regional interconnectedness that underpinned the rapid expansion of nineteenth-century New York.

38

David E. Nye, *American technological sublime* (Cambridge, Mass.: MIT Press, 1994), p. 1. On the iconography of nineteenth-century water engineering and changing conceptions of American landscape, see Patrick McGreevy, "Imagining the future at Niagara Falls," *Annals of the Association of American Geographers* 77 (1987): 48–62, and William Irwin, *The new Niagara: tourism, technology, and the landscape of Niagara Falls, 1776–1917* (University Park: Pennsylvania State University Press, 1996). On the cultural history of American technology see also Joseph J. Corn, ed., *Imagining tomorrow: history, technology and the American future* (Cambridge, Mass.: MIT Press, 1986), and Leo Marx, *The pilot and the passenger: essays on literature, technology and culture in the United States* (Oxford: Oxford University Press, 1988).

39

See Laura Vookles Hardin, "Celebrating the aqueduct: pastoral and urban ideals," in Hudson River Museum of Westchester, *The Old Croton Aqueduct,* pp. 49–56. See also Marvin Fisher, *Workshops in the wilderness: the European response to American industrialization, 1830–1860* (New York: Oxford University Press, 1967); John F. Kasson, *Civilizing the machine: technology and republican values in America, 1776–1900* (New York: Grossman, 1976); and Carroll Pursell, *The machine in America: a social history of technology* (Baltimore: Johns Hopkins University Press, 1995).

40

George Templeton Strong, cited in Gerard Koeppel, "A struggle for water," *Invention and Technology* 9 (1994): 18–31.

41

See Joanne Abel Goldman, *Building New York's sewers: developing mechanisms of urban management* (West Lafayette, Ind.: Purdue University Press, 1997), p. 51.

42

See Moehring, *Public works and the patterns of urban real estate growth in Manhattan.*

43

Sam Bass Warner makes a similar observation with respect to nineteenth-century Philadelphia: "Fears of epidemics had created the water system, but once this fear had abated, little or no public support remained to bring the benefits of the new technology to those who could not afford them." Warner, *The private city,* p. 3. On the significance of municipal achievement for the legitimacy of urban government, see Jon C. Teaford, *The unheralded triumph: city government in America, 1870–1900* (Baltimore: Johns Hopkins University Press, 1984), and Raymond Mohl, *The making of urban America* (Wilmington, Del.: Scholarly Resources, 1988). The fetishistic nature of urban infrastructure as a means to mask unchanged social relations is tackled in Maria Kaïka and Erik Swyngedouw, "Fetishising the modern city: the phantasmagoria of urban technological networks," *International Journal of Urban and Regional Research* 24 (2000): 120–138. On the politics of public health reform in American cities see also Barbara Gutmann Rosenkrantz, *Public health and the state: changing views in Massachusetts, 1842–1936* (Cambridge, Mass.: Harvard University Press, 1972); Judith Walzer Leavitt, *The healthiest city: Milwaukee and the politics of health reform* (Princeton, N.J.: Princeton University Press, 1982); and Stuart Galishoof, *Newark: the nation's unhealthiest city, 1832–1895* (New Brunswick, N.J.: Rutgers University Press, 1988).

44

See André Guillerme, "Water for the city," *Rassegna: Themes in Architecture* 57 (1994): 6–21.

45

Joel Shew, *The water-cure manual* (New York: La Morte Barney, 1847), p. 2. See also John G. Coffin, *Discourses on cold and warm bathing; with remarks on the effects of drinking cold water in warm weather,* 2d ed. (Boston: Cummings and Hilliard, 1826).

46

Sigfried Giedion, *Mechanization takes command: a contribution to anonymous history* (New York: Oxford University Press, 1948), p. 684.

47

See Richard L. Bushman and Claudia L. Bushman, "The early history of cleanliness in America," *Journal of American History* 74 (1988): 1213–1238; Maureen Ogle, "Domestic reform and American household plumbing, 1840–1870," *Winterthur Portfolio* 28 (1993): 33–58; Maureen Ogle, *All the modern conveniences: American household plumbing, 1840–1890* (Baltimore: Johns Hopkins University Press, 1996); Richard L. Bushman, *The refinement of America: persons, houses, cities* (New York: Knopf, 1992); Suellen Hoy, *Chasing dirt: the American pursuit of cleanliness* (New York: Oxford University Press, 1995).

48

Alain Corbin, *The foul and the fragrant: odor and the French social imagination* (1982; Cambridge, Mass.: Harvard University Press, 1986). See also Pierre Bourdieu, *Distinction: a social critique of the judgement of taste* (1979; London: Routledge, 1986).

49

See, for example, Lawrence Wright, *Clean and decent: the fascinating history of the bathroom and the water closet* (New York: Viking Press, 1960); Stuart M. Blumin, *The emergence of the middle class: social experience in the American city, 1760–1900* (Cambridge: Cambridge University Press, 1989); Nancy Tomes, "The private side of public health: sanitary science, domestic hygiene and the germ theory, 1870–1900," *Bulletin of the History of Medicine* 64 (1990): 467–480.

50

William Paul Gerhard, *The drainage of a house* (Boston: Rand Avery, 1888); Giedion, *Mechanization takes command;* New York City Committee of Seventy, *Preliminary report of the sub-committee on baths and lavatories,* 1895. Although water became a fashionable cultural commodity, the adoption of bathrooms in every home did not occur until the twentieth century. The standard bathroom design, for example, did not emerge until the mass production of enameled fixtures in the 1920s and the spread of the "American Compact Bathroom" with its origins in hotel architecture. See also Susan J. Kleinberg, "Technology and women's work: the lives of working class women in Pittsburgh, 1870–1900," in Martha Moore Trescott, ed., *Dynamos and virgins revisited*

(Metuchen, N.J.: Scarecrow Press, 1979), pp. 185–204; John Duffy, *A history of public health in New York City, 1866–1966* (New York: Russell Sage Foundation, 1974), p. 44; Marilyn Thornton Williams, "New York City's public baths: a case study in urban progressive reform," *Journal of Urban History* 7 (1980): 49–81.

51

See Anthony Jackson, *A place called home: a history of low-cost housing in Manhattan* (Cambridge, Mass.: MIT Press, 1976); David Glassberg, "The public bath movement in America," *American Studies* 20 (1979): 5–21; Ogle, "Water supply, waste disposal, and the culture of privatism"; Richard Plunz, *A history of housing in New York City: dwelling type and social change in the American metropolis* (New York: Columbia University Press, 1990).

52

George E. Waring, "The disposal of a city's waste," *North American Review* (July 1895), p. 49.

53

New York City Metropolitan Sewerage Commission, *Sewerage and sewage disposal in the metropolitan district of New York and New Jersey* (New York: Martin B. Brown, 1910).

54

The sanitary inspector Robert Newman, cited in James E. Serrell, *Compilation of facts representing the present condition of the sewers and their deposits in the City of New York* (New York: Bergen & Tripp, 1866), p. 10.

55

Gail Caskey Winkler and Charles E. Fischer III, *The well-appointed bath: authentic plans and fixtures from the early 1900s* (Washington, D.C.: Preservation Press/National Trust for Historic Preservation, 1989).

56

John R. Freeman, *Report upon New York's water supply with particular reference to the need of procuring additional sources and their probable cost with works constructed under municipal ownership* (New York: Martin B. Brown, 1900). New York's ascendancy was clearly short-lived: by 1855, for example, New York had only 1,361 baths for its population of 629,904 compared with 3,521 for a population of nearly 340,000 in Philadelphia. See Edgar W. Martin, *The standard of living in 1860* (Chicago: University of Chicago Press, 1942).

57

See Bushman and Bushman, "The early history of cleanliness in America"; Ellen Lupton and J. Abbott Miller, *The bathroom, the kitchen and the aesthetics of waste: a process of elimination* (New York: Kiosk, 1992).

58

Weidner, *Water for a city,* p. 115. See also Duffy, *A history of public health in New York City, 1866–1966,* p. 77.

59

James C. Bayles, *Pipe gallery experience* (New York: Martin B. Brown, 1903), p. 3.

60

Elisha Harris, *Cholera prevention: examples and practice and a note on the present aspects of the epidemic* (New York: Appleton & Company, 1867).

61

Gerald R. Iwan, "Drinking water quality concerns of New York City, past and present," *Annals of the New York Academy of Sciences* 502 (1987): 183–204. The first water quality laboratories for an American city had been established in Boston in 1889. On the history of water pollution in upstate New York see also G. Harris and S. Wilson, "Water pollution in the Adirondack Mountains: scientific research and governmental response, 1890–1930," *Environmental History Review* 17 (1993): 47–81.

62

Arnold H. Ruge to William L. Strong, "Rational improvements to purify the water supply of the city of New York" (17 December 1894). Papers held by the New-York Historical Society.

63

Other changes include a reforestation program and improved policing of the city's watershed. See Frank E. Hale, "The development of the sanitary safeguards of the New York City water supply and its relation to typhoid fever and other diseases," *Municipal Engineers Journal* 7 (1921): 30–61. As a result of these different initiatives, typhoid death rates in Manhattan fell from around 20 per 100,000 population in 1901 to under 2 by 1919.

64

On the emergence of urban planning in American cities see, for example, M. Christine Boyer, *Dreaming the rational city: the myth of American city planning* (Cambridge, Mass.: MIT Press, 1983); John D. Fairfield, "The scientific management of urban space: professional city planning and the legacy of progressive reform," *Journal of Urban History* 20 (1994): 179–204; Nelson P. Lewis, *The planning of the modern city* (New York: Wiley, 1916); Stanley K. Schultz and Clay McShane, "To engineer the metropolis: sewers, sanitation, and city planning in late nineteenth-century America," *Journal of American History* 65 (1978): 389–411; Joel A. Tarr, *The search for the ultimate sink: urban pollution in historical perspective* (Akron, Ohio: University of Akron Press, 1996).

65

George T. Hammond, "City planning and the engineer," *Municipal Engineers Journal* 2 (1916): 368, 338.

66

Mike Davis, *Prisoners of the American dream: politics and economy in the history of the US working class* (London: Verso, 1986), p. 120. See also Andrew Jamison, "American anxieties: technology and the reshaping of republican values," in Mikael Hård and Andrew Jamison, eds., *The intellectual appropriation of technology: discourses on modernity, 1900–1939* (Cambridge, Mass.: MIT Press, 1998), pp. 69–100.

67

On the evolution and diversity of American engineering traditions see David Hovey Calhoun, *The American civil engineer: origins and conflict* (Cambridge, Mass.: MIT Press, 1960); Monte A. Calvert, *The mechanical engineer in America, 1830–1910* (Baltimore: Johns Hopkins University Press, 1967); R. S. Kirby, S. Withington, A. B. Darling, and F. G. Kilfour, *Engineering in history* (New York: McGraw-Hill, 1956); Donald C. Jackson, "Engineering in the progressive era: a new look at Frederick Haynes Newell and the US Reclamation Service," *Technology and Culture* 34 (1993): 539–574; Peter Meiksins, "The 'revolt of the engineers' reconsidered," in Terry S. Reynolds, ed., *The engineer in America: a historical anthology from Technology and Culture* (Chicago: University of Chicago Press, 1991), pp. 399–426; R. H. Merritt, *Engineering in American society, 1850–1875* (Lexington: University of Kentucky Press, 1969); and Tom F. Peters, *Building the nineteenth century* (Cambridge, Mass.: MIT Press, 1996).

68

The tension between technical expertise and political power can be illustrated by the evolving relationship between New York City government and New York State-based interests. In the period between 1849 and 1857, for example, a series of city charter revisions undertaken by the State Legislature sought to strengthen the power of technical elites over newly emerging city-based political machines. The 1857 charter revision was especially significant in increasing the role of engineers and technical experts in urban management. In 1870, however, a new city charter firmly reasserted the power of city-based political interests and established a powerful new Department of Public Works which was to become a pivotal focus of power and patronage in the Tweed era of New York City governance. In the wake of the 1873 financial panic there was a renewed emphasis on investment in public works, and the transformation of New York's physical environment gathered pace. See Goldman, *Building New York's sewers;* Moehring, *Public works and the patterns of urban real estate growth in Manhattan,* p. 39; Schultz and

McShane, "To engineer the metropolis." As the cities grew, public contracts multiplied for water, gas, electricity, street railroads, and telephones, which helps to explain the rationale of greater autonomy of civil institutions from the patronage of local government. See Boyer, *Dreaming the rational city;* Sarah S. Elkind, "Building a better jungle: anti-urban sentiment, public works, and political reform in American cities, 1880–1930," *Journal of Urban History* 24 (1997): 53–88.

69

Access to Croton water had proved a significant pretext for these outer boroughs' voting to join New York City. The water shortages facing Brooklyn in the late nineteenth century provided further impetus for the expansion of the Croton system and a new emphasis on regional water resources planning. See Brooklyn Division of Water Supply and I. M. deVarona, *History and description of the water supply of the city of Brooklyn* (Brooklyn, N.Y.: Brooklyn Commissioner of City Works, 1896); Theodore Weston, *A report on the extent and character of the district supplying water to the City of Brooklyn* (Brooklyn, N.Y.: D. Van Nostrand, 1861). A systematic analysis of the city's water needs was developed by 1903 by the Hering-Burr-Freeman Commission, which laid the scientific basis for the expansion of the water system into the Catskill Mountains. The political impetus for the completion of the project was also bolstered by the drought of 1910, the worst since 1826.

70

See Board of Water Supply of the City of New York, *A general description of the Catskill water supply and of the project for an additional supply from the Delaware River watershed and the Rondout Creek* (1940); J. Paul Sehgal, "New York's water problems," *Municipal Engineers Journal* 79 (1991): 24–34.

71

Gilbert J. Fowler, cited in Metropolitan Sewerage Commission of New York, *Preliminary reports on the disposal of New York's Sewage VII: Critical reports of Dr Gilbert J. Fowler of Manchester, England, and Mr John D. Watson of Birmingham, England, on the projects of the metropolitan sewerage commission with special reference to the plans proposed for the Lower Hudson, Lower East River and Bay Division* (1913), p. 33.

72

Weidner, *Water for a city,* p. 140.

73

See Silas B. Dutcher to Bird S. Coler, 5 August 1899, papers held by the New-York Historical Society. This letter reveals the failure of the Ramapo Water Company to divulge requested technical and financial information to the city's comptroller.

74

T. Hochlerner, "Distribution system for New York," *Water Works Engineering* 95 (1942): 1262–1266.

75

See, for example, William L. Kahrl, *Water and power: the conflict over Los Angeles' water supply in the Owens Valley* (Berkeley: University of California Press, 1982); R. Lowitt, "The Hetch-Hetchy controversy," *California History* 74 (1995): 190–203; Robert A. Sauder, *The lost frontier: water diversion and the destruction of Owens Valley agriculture* (Tuscon: University of Arizona Press, 1993); John Walton, *Western times and water wars: state, culture, and rebellion in California* (Berkeley: University of California Press, 1992); Weidner, *Water for a city*, p. 231.

76

Weidner, *Water for a city*, pp. 241, 263.

77

Ibid., pp. 191, 272.

78

Ibid., p. 271. See also E. J. Clark, "The New York City water shortage," *Municipal Engineers Journal* 36 (1950): 92–104; Anastasia Van Burkalow, "The geography of New York City's water supply: a study of interactions," *Geographical Review* 49 (1959): 369–386.

79

New York Herald, cited in Weidner, *Water for a city*, p. 271. The commemorative medal of the Mayor's Catskill Celebration Committee described the new water system as "an achievement of civic spirit, scientific genius, and faithful labor." See Edward E. Hall, *Water for New York City* (1917; Saugerties, N.Y.: Hope Farm Press, 1993), p. 122.

80

Weidner, *Water for a city*, p. 290.

81

Board of Water Supply of the City of New York, *A general description of the Catskill water supply*, 1940, papers held by the New-York Historical Society.

82

Charles M. Clark, "Development of the Catskill supply," *Water Works Engineering* 95 (1942): 1268–1269; Charles M. Clark, "History of the Delaware supply," *Water Works Engineering* 95 (1942): 1276–1277; Fiorello La Guardia, "A Letter from Mayor La Guardia," *Water Works Engineering* 95 (1942): 1256.

83

Rexford G. Tugwell, "San Francisco as seen from New York," *National Conference on Planning* (proceedings of the conference held at San Francisco, 8–11 July 1940), pp. 182–188. See also Thomas Kessner, *Fiorello H. La Guardia and the making of modern New York* (New York: McGraw-Hill, 1989); William Stanley Parker, "Public works: the future share of federal and non-federal agencies," *National Conference on Planning* (proceedings of the conference held at Boston, Massachusetts, 15–17 May 1939).

84

"New York City's gigantic thirst," *New York Herald Tribune,* 15 August 1959.

85

Edward J. Clark, "New York City water supply system, unusual occurrences and unique problems," *Municipal Engineers Journal* 34 (1948): 103–115.

86

"Mayor asks curb on use of water," *New York Times,* 15 October 1963.

87

Abel Wolman, "The metabolism of cities," *Scientific American* 213 (1965): 180.

88

Anthony Vidler, "Dark space," in *The architectural uncanny: essays in the modern unhomely* (Cambridge, Mass.: MIT Press, 1992), p. 167. On the "hidden city" see, for example, "Many in Manhattan fear man-made menace lurks underground," *Washington Post,* 16 November 1989. On the underground city see also Robert Daley, *The world beneath the city* (Philadelphia: Lippincott, 1959); Harry Granick, *Underneath New York,* rpt. with an introduction by Robert E. Sullivan, Jr. (1947; New York: Fordham University Press, 1991); Stanley Greenberg, *Invisible New York: the hidden infrastructure of the city* (Baltimore: Johns Hopkins University Press, 1998); Peter Seidel, *Unterwelten: Orte im verborgenen/sites of concealment,* texts by Manfred Sack and Klaus Kemp (Tübingen: Ernst Wasmuth Verlag, 1993); Jennifer Toth, *The mole people: life in the tunnels beneath New York City* (Chicago: Chicago Review Press, 1993); and Rosalind H. Williams, *Notes on the underground: an essay on technology, society, and the imagination* (Cambridge, Mass.: MIT Press, 1990).

89

Erik Swyngedouw, "The city as a hybrid: on nature, society and cyborg urbanization," *Capitalism, Nature, Socialism* 7 (1997): 65–80.

90

Gottmann, *Megalopolis,* p. 3. See also Kenneth T. Jackson, "The capital of capitalism: the New York metropolitan region, 1890–1940," in Anthony Sutcliffe, ed., *Metropolis, 1890–1940*

(Chicago: University of Chicago Press, 1984); Cleveland Rogers and Rebecca C. Rankin, *New York: the world's capital city* (New York: Harper & Brothers, 1948); M. Shefter, ed., *Capital of the American century: the national and international influence of New York City* (London: Sage, 1983).

91

Lisa Gitelman, "Negotiating a vocabulary for urban infrastructure, or, the WPA meets the Teenage Mutant Ninja Turtles," *Journal of American Studies* 26 (1992): 147–158; Thomas P. Hughes, *American genesis: a century of invention and technological enthusiasm, 1870–1970* (New York: Viking, 1989).

92

Sidney Hornstein, American Museum of Natural History, interview with the author (17 April 1996). A compendium of urban ruins is contained in Camilo José Vergara, *The new American ghetto* (New Brunswick, N.J.: Rutgers University Press, 1995).

93

Francis X. McCardle, "Water supply for New York City in the 1980s," *Journal of the American Water Works Association* (March 1982), p. 139.

94

For overviews of the causes and consequences of New York's fiscal crisis see R. Alcaly and D. Mermelstein, eds., *The fiscal crisis of American cities: essays on the political economy of urban America with special reference to New York* (New York: Vintage Books, 1977); R. W. Bailey, *The crisis regime: the New York City financial crisis* (Albany: State University of New York Press, 1984); Ester R. Fuchs, *Mayors and money: fiscal policy in New York and Chicago* (Chicago: University of Chicago Press, 1992); E. Lichten, *Class, power and austerity: the New York City fiscal crisis* (South Hadley, Mass.: Bergin and Garvey, 1986); William K. Tabb, *The long default: New York City and the urban fiscal crisis* (New York: Monthly Review Press, 1982).

95

The city's crisis in the 1970s coincided with the development of neo-Marxian perspectives on cities and urban planning by Manuel Castells, David Harvey, and Henri Lefebvre. Three critical texts from this era are Manuel Castells, *The urban question: a Marxist approach* (1972; Cambridge, Mass.: MIT Press, 1977); David Harvey, *Social justice and the city* (London: Edward Arnold, 1973); and Henri Lefebvre, *The production of space,* trans. D. Nicholson-Smith (1974; Oxford and Cambridge, Mass.: Blackwell, 1991).

96

See David Harvey, *The limits to capital* (Oxford: Blackwell, 1982); David Harvey, *The urbanization of capital: studies in the history and theory of capitalist urbanization 2* (Baltimore: Johns Hopkins

University Press, 1985). On capital switching in urban space see also Robert A. Beauregard, "Capital restructuring and the new built environment of global cities: New York and Los Angeles," *International Journal of Urban and Regional Research* 15 (1991): 90–105; Robert A. Beauregard, "Capital switching and the built environment: United States, 1970–1989," *Environment and planning A* 26 (1994): 715–732; and Ernest Mandel, *Late Capitalism,* trans. Joris De Bres (1972; London: Verso, 1978).

97

D. A. Grossman, "Debt and capital management," in C. Brecher and R. D. Horton, eds., *Setting municipal priorities, 1983* (London and New York: New York University Press, 1982), pp. 120–151; J. A. Hartman, "Capital resources," in C. Brecher and R. D. Horton, eds., *Setting municipal priorities, 1986* (London and New York: New York University Press, 1985), pp. 139–169.

98

Mike Davis, "Who killed Los Angeles? Part One," *New Left Review* 197 (1993): 3–28.

99

City of New York Office of the Comptroller, *Dilemma in the millennium: capital needs of the world's capital city* (New York: Office of the Comptroller, 1998); Cooper Union for the Advancement of Science and Art, *The age of New York City infrastructure* (New York: CUASA, 1991). David A. Golub, Press Office, New York City Department of Environmental Protection, interview with the author (7 July 1995); Frank Oliveri, Senior Manager in the Bureau of Waste Water Treatment, New York City Department of Environmental Protection, interview with the author (7 July 1995); Clark Wieman, Senior Research Fellow, Cooper Union for the Advancement of Science and the Arts, New York, interview with the author (22 March 1995).

100

High-profile disasters include: "Flood in the streets: 136-year-old Brooklyn water main bursts," *New York Newsday,* 22 January 1994; "Rupture of pipe floods subway in Manhattan," *New York Times,* 10 July 1995; "Water main break leads to evacuation," *New York Times,* 2 June 1999. The public health issues associated with underinvestment in water supply infrastructure are examined in Natural Resources Defense Council, *Victorian water treatment enters the 21st century: public health threats from water utilities' ancient treatment and distribution systems* (New York: Natural Resources Defense Council, 1994).

101

Mark Izeman, Project Attorney, Natural Resources Defense Council, New York, interview with the author (26 June 1995); Kemba Johnson, "H_2Owe," *City Limits* (October 1997), pp. 10–11; New York City Office of the Comptroller, *Who's accountable for soaring water and sewer fees?* (New York: Office of the Comptroller, 1991; Brian Purlee and Henry Smeal, *The future cost of water in New York City* (New York: Citizens Budget Commission, 1994).

102

Alex Ulam, "Shadow of a drought," *City Limits* (September/October 1999), p. 10.

103

On the Koch era of public-sector restructuring see John H. Mollenkopf, *A phoenix in the ashes: the rise and fall of the Koch coalition in New York City politics* (Princeton, N.J.: Princeton University Press, 1992). On the broader political and economic context for public-sector restructuring in the United States see M. Aglietta, *A theory of capitalist regulation: the US experience* (1979; London: Verso, 1987).

104

Johnson, "H_2Owe."

105

The marked fluctuations in water consumption within the overall trend illustrated in figure 1.11 can be attributed to the impact of increasingly frequent droughts, notably in 1961–1967, 1980–1981, 1985, 1989, and 1993. See R. Cropf, "Water resources," in C. Brecher and R. D. Horton, eds., *Setting municipal priorities, 1990* (New York: New York University Press, 1989), pp. 173–197. In the late 1980s the city began to introduce universal metering, involving a computerized automated reading and billing system, along with a range of proposals to promote water conservation such as the installation of water-saving plumbing devices, modifications to building specifications, and water audits for large users. As a consequence of the stabilization of growth in demand for water, current policy debate over the capacity of the water supply system may well shift toward the long-term management of climatic uncertainty.

106

"Firms vying for water bond deal," *New York Newsday,* 9 May 1995.

107

The comptroller is the most senior financial post in the New York City administration and is responsible for auditing all financial transactions. See "Comptroller threatens to stop water system sale," *New York Times,* 24 June 1995; "Mayor blocked on plan to sell water system," *New York Times,* 28 June 1995; "Mr. Hevesi's timely warning," *New York Times,* 29 June 1995.

108

"Top court sinks city water sale," *New York Newsday,* 21 March 1997. The case was ultimately won on the grounds that it is illegal to divert funds from the sale of water assets into other unrelated purposes. Nancy Anderson, Senior Environmental Policy Advisor, Office of Policy Management, City of New York, interview with the author (16 September 1999).

109

Thomas H. Garver, "Serving places," in Greenberg, *Invisible New York,* p. 4.

110

James C. de Haven and Jerome W. Milliman, *Water supply: economics, technology, and policy* (Chicago: University of Chicago Press, 1960).

111

"It's wet, free and gets no respect," *New York Times,* 11 November 1991.

112

Letter to the editor from Assemblyman John Ravitz, *New York Times,* 30 November 1991.

113

New York City Department of Environmental Protection, *Review of Croton system water quality, and current and future water quality standards* (New York: Department of Environmental Protection, 1994).

114

New York City Department of Environmental Protection, *Range and average of drinking water characteristics of the Catskill/Delaware and Croton supplies from respective points in the New York City distribution system for the year 1993* (New York: Department of Environmental Protection, 1995). The technical aspects to the water quality debate are focused on a series of indicators, principally the concentration of the *Escherichia coli* bacterium (a key fecal coliform indicating contamination with sewage and the possible presence of other more dangerous pathogens such as hepatitis A, intestinal parasites, and *Salmonella*), the level of turbidity (cloudiness from particulate matter), the level of trihalomethanes (by-products of chlorine interacting with organic matter during the disinfection process), and the presence of pathogens, notably *Cryptosporidium, Giardia,* and enteric viruses.

115

"Detection of *E. coli* infections on the rise," *New York Times,* 13 June 1995; "An outbreak under wraps," *New York Newsday,* 19 June 1996.

116

"How safe is our water?," *New York* (16 January 1995).

117

A contributory factor in the breakdown of trust between the public and regulatory agencies has been a series of testing scandals in the mid 1990s in which former city employees allege that levels of *E. coli* and other contaminants were deliberately altered in test results to avoid a public health crisis. See "An outbreak under wraps," *New York Newsday,* 19 June 1995.

118

"Some bottled water is called unsafe," *New York Times,* 31 March 1999.

119

Filtration of the Croton system has begun with a pilot filtration plant at the Jerome Park Reservoir (the first water filtration ever undertaken by the city), and a full-scale filtration plant located at Mosholu Golf Course in the Bronx is expected to be completed by the year 2007 at a cost of around $660 million. The supply of inferior Croton water to poorer parts of the city has led to allegations of environmental racism and also to "water quality alliances" between the environmental justice movement and the Hospital Workers Union, yet Croton water is also supplied to the affluent Upper East Side and parts of Riverside Drive.

120

The presence of *Cryptosporidium* and other pathogens has also caused alarm in New York's gay community since some of these organisms are known to cause potentially fatal illnesses in immuno-supressed individuals; they also present a threat to children and the elderly. "Lethal liquid? Parasites in our water," *New York Newsday,* 21 June 1995.

121

See, for example, L. O. Gostin, Z. Lazzarini, V. S. Neslund, and M. T. Ostenholm, "Water quality laws and waterborne diseases: *Cryptosporidium* and other emerging pathogens," *American Journal of Public Health* 90 (2000): 847–853; T. S. Steiner, N. M. Thielman, and R. L. Guerrant, "Protozoal agents: what are the dangers for the public water supply?," *Annual Review of Medicine* 48 (1997): 329–340; R. L. Berkelman, "Emerging infectious diseases in the United States, 1993," *Journal of Infectious Diseases* 170 (1994): 272–277; "Water: danger on tap," *New York Daily News,* 27 July 1997. The use of sophisticated filtration technologies does not guarantee water quality: in Milwaukee, Wisconsin, the presence of *Cryptosporidium* in a filtered public water supply caused more than 400,000 people to become ill and led to the death of over a hundred mostly elderly or immuno-suppressed people. For greater detail on cryptosporidiosis, see M. W. LeChevallier, W. D. Norton, and R. G. Lee, "Occurrence of *Giardia* and *Cryptosporidium* spp. in surface water supplies," *Applied and Environmental Microbiology* 57 (1991): 2610–2616; and W. R. MacKenzie, "A massive outbreak in Milwaukee of *Cryptosporidium* infection transmitted through the public water supply," *New England Journal of Medicine* 331 (1994): 161–167.

122

A paradoxical outcome of the tightening of federal water quality standards for New York is that fiscal pressures are forcing a tradeoff between different environmental goals. As Edward Scheader notes, a federal Surface Water Treatment Rule primarily concerned with biological safety compels improving disinfection efficiency and increasing chlorine levels; a Lead and Copper Rule compels increasing water pH to reduce corrosivity, a step that will have the two-fold effect of increasing trihalomethane production and at the same time reducing disinfection efficiency; while a Disinfection By-Products Rule will severely limit the use of chlorine as a primary disinfecting agent. See Edward C. Scheader, "The New York City water supply: past, present & future," *Civil Engineering Practice* (Fall 1991), pp. 7–20. The feasibility of the city's goal of avoiding filtration for the Catskill-Delaware system has recently been subjected to intense scrutiny in a report prepared by the National Academy of Sciences on behalf of City Comptroller Alan G. Hevesi. The report called for tighter controls on the sources of cryptosporidiosis, *Giardia*, and organic compounds in the water supply, as well as the modernization of septic systems across the watershed. A priority identified for the city is preventing polluted runoff from reaching the Kensico Reservoir north of White Plains, which serves as a holding basin for all of the upstate water system. Nancy Anderson, interview with the author (16 September 1999).

123

Simon Gruber, Gaia Institute of the Cathedral of St. John the Divine, interview with the author (21 March 1995).

124

"Water projects' bill worries even environmentalists," *New York Times*, 15 June 1998. There is extensive evidence that the fiscal burden of ecological modernization falls disproportionately on lower-income groups. The modernization of water quality infrastructure is an integral element in this market-led technological paradigm for water management.

125

Recent census data reveal that there are actually fewer people now living in much of the Catskill-Delaware water catchment area than in 1860. Nancy Anderson, interview with the author (16 September 1999).

126

Simon Gruber, interview with the author (21 March 1995).

127

For greater detail on development pressures in the New York metropolitan region see Robert D. Yaro and Tony Hiss, *A region at risk: the third regional plan for the New York-New Jersey-Connecticut metropolitan area* (Washington, D.C., and Covelo, Calif.: Island Press, 1996).

128

"Governor Pataki's watershed," *New York Times,* 26 June 1995.

129

Iwan, "Drinking water concerns of New York City, past and present," p. 203.

130

Robert F. Kennedy, Jr., *Cops in cuffs: the failure of environmental enforcement and security in the New York City watershed* (Garrison, N.Y.: Hudson Riverkeeper Fund, 1999); "Rudy's letting city's reservoirs go down the drain," *New York Daily News,* 23 October 1994.

131

"Culture of mismanagement: environmental protection and enforcement at the New York City Department of Environmental Protection," *Pace Environmental Law Review* (1997), 15.

132

On the case of Boston, which shares some similarities with that of New York City, see "EPA files suit to force State to build filtration plant for Hub-area water," *Boston Globe,* 13 February 1998; "Filtration fight brews in Boston," *Engineering News Record,* 22 December 1997; Philip E. Steinberg and George E. Clark, "Troubled water? Acquiescence, conflict, and the politics of place in watershed management," *Political Geography* 18 (1999): 477–508.

133

Simon Gruber, interview with the author (24 April 1995); "Reservoir towns and New York City reach agreement," *New York Times,* 27 October 1995; "At last, a watershed agreement," *New York Times,* 3 November 1995. Developments in New York are not without precedent elsewhere in the United States: California's Delta-Bay Accord of 1994 is another important example of new urban-rural policy fora. See Eddie Nickens, "A watershed paradox," *American Forests* 103 (1998): 21–24.

134

Simon Gruber, interview with the author (15 June 1995). Additional data taken from field notes at the meeting of the Ad Hoc Forestry Task Force held in Liberty, New York, 22 June 1995. An important difficulty in watershed protection regulation is the lack of basic wastewater infrastructure in many of the more remote upstate communities. Many small low-income towns in the Catskill-Delaware system such as Bedford and Katanah have no central sewer. See J. Prendergast, "Small systems struggle," *Civil Engineering* 62 (1992): 36–37. In the Catskill-Delaware watershed, 21 of the 38 towns lack zoning ordinances, the most rudimentary form of development control. See also D. K. Gordon and R. F. Kennedy, *The legend of city water: recommendations for rescuing the New York City water supply* (Garrison, N.Y.: Hudson Riverkeeper Fund, 1991); Wayne Barrett, "Watershed waffle: Rudy takes a dive on city wetlands for upstate GOP," *Village Voice* (30 March 1999), p. 23; "Doing battle on stiff rules for watershed," *New York Times,* 27 July 1998; "City reverses itself on effort to protect land at reservoirs," *New York Times,* 19 March 1999; "US says New York City may have to spend $6 billion on filtration," *New York Times,* 1 June 2000. On the history of land use conflict in upstate New York see also Karl Jacoby, "Class and environmental history: lessons from the 'War in the Adirondacks,'" *Environmental History* 2 (1997): 324–342.

135

Development pressures in New York's watershed are most acute in Westchester and Putnam counties east of the Hudson River. Nancy Anderson, personal communication with the author (8 October 1999).

136

Lewis Mumford, "The sky line: the architecture of power," *New Yorker* (7 June 1941), pp. 58–60. See also Aaron Betsky, "Measured immensity: Hoover Dam at fifty," *Progressive Architecture* 66 (1985): 38; Brian Black, "Authority in the valley: TVA in Wild River and the popular media, 1930–1940," *Journal of American Culture* 18 (1995): 1–14; Margot W. Garcia, "An everlasting monument: the building of Roosevelt Dam," *Triglyph: A Southwestern Journal of Architecture and Environmental Design* 5 (1993): 34–46; Jean-François Lejeune, "Democratic pyramids: the works of the Tennessee Valley Authority," *Rassegna: Themes in Architecture* 63 (1995): 46–57; Marian Moffett and Lawrence Wodehouse, "Noble structures set in handsome parks: public architecture of the TVA," *Modulus: University of Virginia Architectural Review* 17 (1984): 74–83; Richard Guy Wilson, "Massive deco monument: the enduring strength of Boulder (Hoover) Dam," *AIA: Architecture* (December 1983), pp. 45–47; Richard Guy Wilson, "Machine-age iconography in the American West: the design of Hoover Dam," *Pacific Historical Review* 54 (1985): 463–497.

137

See Susan Fainstein and Norman Fainstein, "Technology, the new international division of labor, and location: continuities and disjunctures," in Robert Beauregard, ed., *Economic restructuring and political response* (Newbury Park, Calif.: Sage Publications, 1989).

138

"How to sell the world's water industry," *Financial Times,* 2 October 1997.

139

"Controversy hits Manila water privatisation plan," *Financial Times,* 17 January 1997; "Water: trouble in the pipeline," *Financial Times,* 14 September 1998; "Doubts over Brazilian water privatisation," *Financial Times,* 13 November 1998; "Anger in the Andes," *Financial Times,* 4 July 2000; David Saurí and Francisco Manuel Muñoz, "The limits of ecological modernization in the multiplied city: equity and conflict over water costs in the metropolitan area of Barcelona," in Erik Swyngedouw, Leandro del Moral, and Grigoris Kafkalas, eds., *Sustainability, risk and nature: the political ecology of water in advanced societies* (Oxford: Oxford Centre for Water Research, 1999), pp. 195–201.

140

See M. J. Pfeffer and J. M. Stycos, *Watershed views: a public opinion survey on the New York City watershed* (Ithaca, N.Y.: Department of Rural Sociology and Population and Development Program, Cornell University, 1994); K. A. Stave, "Resource conflict in the New York City Catskill watersheds: a case for expanding the scope of water resource management," in L. H. Austin, ed., *Water in the 21st century: conservation, demand and supply* (Herndon, Va.: Proceedings of the American Water Works Association Annual Spring Symposium, 1995), pp. 61–68.

141

See David C. Major, "Urban water supply and global environmental change: the water supply system of New York City," in R. Herrmann, ed., *Proceedings of the 28th annual American Water Resources Association conference and symposium: managing resources during global change* (Bethesda, Md.: American Water Resources Association, 1992), pp. 377–385; "Getting ready for a hotter, wetter future," *New York Times,* 1 December 1997; "Report warns New York of perils of global warming," *New York Times,* 30 June 1999; "Water supply also affected by global warming," *Village Voice* (3 August 1993).

142

Arjun Appadurai, *Modernity at large: cultural dimensions of globalization* (Minneapolis: University of Minnesota Press, 1996), p. 19. For starkly different views of the relationship between politics and modernity see Jürgen Habermas, *The philosophical discourse of modernity: twelve lectures* (1985; Cambridge: Polity Press, 1987); Eric Hobsbawm, *Age of extremes: the short twentieth century, 1914–1991* (London: Michael Joseph, 1994); and E. M. Wood, *Democracy against capitalism: renewing historical materialism* (Cambridge: Cambridge University Press, 1995).

143
See the critique of Donald Worster's conspiratorial view of urban water demands developed by Douglas Edward Kupel, "Urban water in the arid West: municipal water and sewer utilities in Phoenix, Arizona" (PhD dissertation, Arizona State University, 1995).

2

SYMBOLIC ORDER AND THE URBAN PASTORAL

1
Henry Cleaveland, "The Central Park," *Appleton's Journal* 3 (1870): 671.

2
John W. Reps, *The making of urban america: a history of city planning in the United States* (Princeton, N.J.: Princeton University Press, 1965), p. 336.

3
Frederick Law Olmsted was born in Hartford, Connecticut, in 1822 and died in Belmont, Massachusetts, in 1903. In addition to his work as a landscape architect, Olmsted also managed an experimental farm on Staten Island and during the Civil War was appointed to run the United States Sanitary Commission. For biographical details see Albert Fein, *Frederick Law Olmsted and the American environmental tradition* (New York: George Braziller, 1972); Charles Capen McLaughlin, "Olmsted's odyssey," *Wilson Quarterly* (Summer 1982): 78–87; Laura Wood Roper, *FLO: a biography of Frederick Law Olmsted* (Baltimore: Johns Hopkins University Press, 1973); John Emerson Todd, *Frederick Law Olmsted* (Boston: Twayne Publishers, 1982). From 1858 to 1872 Olmsted worked in partnership with Calvert Vaux. Vaux was born in London in 1824 and died in Brooklyn in 1895. As an architect Vaux played a significant role in popularizing Gothicism and romanticism in North America, but his contribution to the history of landscape design has been relatively neglected. For biographical details see Francis R. Kowsky, *Country, park and city: the architecture and life of Calvert Vaux* (Oxford: Oxford University Press, 1998). For a detailed analysis of Vaux's contribution to Central Park see Roy Rosenzweig and Elizabeth Blackmar, *The park and the people: a history of Central Park* (Ithaca, N.Y.: Cornell University Press, 1992).

4
Given Olmsted's distrust of organized labor it is very difficult to sustain the argument that he was a kind of early socialist reformer, yet both Chadwick and Fein develop this line of argument by

emphasizing the influence on Olmsted of the "social architecture" of Parke Godwin and the American socialist followers of Charles Fourier and Louis Blanc. See George F. Chadwick, *The park and the town: public landscape in the nineteenth and twentieth centuries* (London: Architectural Press, 1965), and Fein, *Frederick Law Olmsted and the American environmental tradition.* Although we know from Olmsted's writings that he was impressed by the Fourierist community at Red Bank, New Jersey, which he visited in 1852, he also noted that "I am not a Fourierist for myself"; quoted in Melvin Kalfus, "Olmsted: a psychohistorical perspective," in Bruce Kelly, Gail Travis Guillet, and Mary Ellen W. Hern, eds., *Art of the Olmsted landscape* (New York: New York City Landmarks Preservation Commission and The Arts Publisher, 1981), p. 144. Olmsted's extensive (and often somewhat discursive) writings suggest that his political outlook combined the prevailing late nineteenth-century ideology of social Darwinism with an ambivalence about the institutional power of capital to shape urban space. He also held classical liberal views with regard to both economic competition and the possibilities for labor organization (a tension that emerged through his managerial role in the construction of Central Park). Above all, Olmsted was more interested in stability than change, hoping that "the tensions of a newly urban nation might be moderated by structural arrangements, both political and aesthetic, to foster respect among rival social groups." See Geoffrey Blodgett, "Frederick Law Olmsted: landscape architecture as conservative reform," *Journal of American History* 62 (1976): 870.

5

See, for example, Albert Fein, "The Olmsted renaissance: a search for national purpose," in Kelly, Guillet, and Hern, eds., *Art of the Olmsted landscape,* p. 99.

6

Spirn advocates a selective appropriation of Olmsted's legacy without his attachment to the moral effects of nature and his endorsement of a technical elite, but she never satisfactorily explores the ambiguity of nature-based conceptions of urban form within the context of capitalist urbanization. See Anne W. Spirn, "Constructing nature: the legacy of Frederick Law Olmsted," in William Cronon, ed., *Uncommon ground: rethinking the human place in nature* (New York: W. W. Norton, 1996), pp. 91–114. A similar perspective is advanced by Paul Bennett, who calls for a return to Olmsted's vision through a combination of ecology, aesthetics, and ethics as part of urban "place restoration." Paul Bennett, "Ecologizing Olmsted," *Landscape Architecture* 88 (June 1998): 52–57. And in New York's third regional plan, published in 1996, Robert Yaro and Tony Hiss proclaim that "Olmsted correctly saw these new park systems as places that could heal most of the ills then facing urban America." Robert D. Yaro and Tony Hiss, *A region at risk: the third regional plan for the New York-New Jersey-Connecticut metropolitan area* (New York: Regional Plan Association; Washington, D.C.: Island Press, 1996), p. 87.

7

The key Lefebvrian distinction for the study of public space is between *representational space* (space appropriated through human use) and *representations of space* (impositions of spatial order). See, for example, Henri Lefebvre, *The production of space,* trans. D. Nicholson-Smith (1974; Oxford and Cambridge, Mass.: Basil Blackwell, 1991), and Henri Lefebvre, *Writings on cities,* ed. and trans. Eleonore Kofman and Elizabeth Lebas (Oxford and Cambridge, Mass.: Basil Blackwell, 1996). See also critical commentaries on his work provided by Edward Dimendberg, "Henri Lefebvre on abstract space," in Andrew Light and Jonathan M. Smith, eds., *Philosophy and geography II: the production of public space* (Lanham, Md.: Rowman and Littlefield, 1998), pp. 17–48; Andrew Merrifield, "Place and space: a Lefebvrian reconciliation," *Transactions of the Institute of British Geographers* 18 (1993): 516–531; Don Mitchell, "The end of public space? People's Park, definitions of the public, and democracy," *Annals of the Association of American Geographers* 85 (1995): 108–133; and Neil Smith, "Antinomies of space and nature in Henri Lefebvre's *The production of space,*" in Light and Smith, eds., *Philosophy and Geography II,* pp. 49–69. In addition to this distinction, we should be aware of a third dimension, that of *spatial practice,* encompassing the material reproduction of society from people to physical infrastructure. Both Edward Dimendberg and Neil Smith highlight a number of interesting anomalies in Lefebvre's theory of the social production of space. In particular they emphasize his weakly developed theory of the social production of nature in distinction to the social production of space.

8

Peter Walker and Melanie Simo argue that Olmsted's legacy serves "as a measure of all subsequent work in landscape architecture." See Peter Walker and Melanie Simo, *Invisible gardens: the search for modernism in the American landscape* (Cambridge, Mass.: MIT Press, 1994), p. 5. This intense focus on Olmsted as the park's designer has led to the relative eclipse of other key figures such as Calvert Vaux, Ignaz Pilat (the former director of the Vienna botanical gardens who oversaw the planting of the new park), George E. Waring (the principal engineer), Andrew Haswell Green (Comptroller of the Central Park Commission from 1857 until 1871), and many others, not to mention the thousands of workers and artisans involved in the creation of the park itself. See Rosenzweig and Blackmar, *The park and the people,* p. 149.

9

Irving D. Fisher, *Frederick Law Olmsted and the city planning movement in the United States* (1976; Ann Arbor, Mich.: UMI Research Press, 1986), pp. 4–5.

10

James Marston Fitch, "Design and designer: 19th century innovation," in Kelly, Guillet, and Hern, eds., *Art of the Olmsted landscape,* p. 74.

11

Calvert Vaux, cited in Rosenzweig and Blackmar, *The park and the people,* p. 122.

12

Rem Koolhaas, *Delirious New York: a retroactive manifesto for Manhattan* (1978; Rotterdam: 010 Publishers, 1994); Edward K. Spann, "'The greatest grid': the New York Plan of 1811," in Daniel Schaffer, ed., *Two centuries of American planning* (Baltimore: Johns Hopkins University Press, 1988), pp. 11–39.

13

See Reps, *The making of urban America;* Chadwick, *The park and the town;* and Rosenzweig and Blackmar, *The park and the people.* For similar tensions over privately owned spaces in London see Henry W. Lawrence, "The greening of the squares of London: transformation of urban landscapes and ideals," *Annals of the Association of American Geographers* 83 (1993): 90–118.

14

In 1852, for example, George Templeton Strong described parts of lower Manhattan as "fermenting, putrefying, and pestilential," a chaotic throng of people and animals that he was eager to escape from as soon as possible. Cited in Bruce Kelly, "Art of the Olmsted landscape," in Kelly, Guillet, and Hern, eds., *Art of the Olmsted landscape,* p. 5. On the extent of urban unrest at this time see Joel Tyler Headley, *The great riots of New York, 1712–1873* (New York: E. B. Treat, 1873).

15

Cited in Ian R. Stewart, "Politics and the park: the fight for Central Park," *New-York Historical Society Quarterly* 61 (1977): 128. In 1844, William Cullen Bryant had written how "Madrid and Mexico City [had] their Alamedes, London its Regent's Park, Paris its Champs Elysees and Vienna its Prater." Cited in Eugene Moehring, *Public works and the patterns of urban real estate growth in Manhattan, 1835–1894* (New York: Arno Press, 1981), p. 269.

16

Cited in Reps, *The making of urban America,* p. 331.

17

See George B. Tatum and Elisabeth B. MacDougall, *Prophet with honor: the career of Andrew Jackson Downing, 1815–1852* (Washington, D.C.: Dumbarton Oaks, 1989).

18

Cited in M. Christine Boyer, *Dreaming the rational city: the myth of American city planning* (Cambridge, Mass.: MIT Press, 1983), p. 39.

19

See Susannah S. Zetzel, "The garden in the machine: the construction of nature in Olmsted's Central Park," *Prospects: An Annual of American Cultural Studies* 14 (1989): 291–339.

20

Rosenzweig and Blackmar, *The park and the people,* pp. 16, 23–24.

21

For greater detail on the cultural significance of nineteenth-century American cemeteries see Thomas Bender, "The 'rural' cemetery movement: urban travail and the appeal of nature," *New England Quarterly* 47 (1974): 196–211; John Dixon Hunt, *Gardens and the picturesque: studies in the history of landscape architecture* (Cambridge, Mass.: MIT Press, 1992); Stanley French, "The cemetery as a cultural institution: the establishment of Mount Auburn and the 'rural cemetery' movement," *American Quarterly* 26 (1974): 37–59; Blanche Linden-Ward, "Strange but genteel pleasure grounds: tourist and leisure uses of nineteenth-century rural cemeteries," in Richard E. Meyer, ed., *Cemeteries and grave markers: voices of American culture* (Ann Arbor: UMI Research Press, 1989), pp. 293–328; David Schuyler, *The new urban landscape: the redefinition of city form in nineteenth-century America* (Baltimore: Johns Hopkins University Press, 1986).

22

Cited in Rosenzweig and Blackmar, *The park and the people,* p. 111. In any case, as Rosenzweig and Blackmar point out, neither public health nor the living conditions of the working class had ever been at the heart of the nineteenth-century park movement (p. 55). The list of businessmen and lawyers who petitioned to have Olmsted as park superintendent in 1857 reveals how the creation of Central Park drew support from the highest echelons of New York's political and economic elite. See Henry Hope Reed, "Central Park: the genius of the place," in Kelly, Guillet, and Hern, eds., *Art of the Olmsted landscape,* pp. 125–131.

23

Frederick Law Olmsted, cited in *Civilizing American cities: a selection of Frederick Law Olmsted's writings on city landscapes,* ed. S. B. Sutton (Cambridge, Mass.: MIT Press, 1971), p. 71.

24

The creation of Union Square Park in lower Manhattan provides an interesting parallel to the experience of Central Park. This park very nearly failed to materialize, yet when it finally opened in 1839 it led to greatly enhanced property values in its immediate vicinity. See Rosalyn

Deutsche, *Evictions: art and spatial politics* (Cambridge, Mass.: MIT Press, 1996), p. 23. The lobbyists for Central Park must have been aware of the success of Union Square Park in raising Manhattan land and property values.

25

Michael A. Mikkelson, "A review of the history of real estate on Manhattan Island," in Real Estate Record Association, *A history of real estate, building and architecture in New York City* (New York: Record and Guide, 1898), p. 156. See also Richard M. Hurd, *Principles of city land values,* 2d ed. (New York: Record and Guide, 1905).

26

Fred. B. Perkins, *The Central Park* (New York: Carleton, 1863), pp. 18–19.

27

Mikkelson, "A review of the history of real estate on Manhattan Island," pp. 1–129; Nelson P. Lewis, *The planning of the modern city* (New York: Wiley, 1916).

28

Rosenzweig and Blackmar, *The park and the people,* p. 18.

29

Ibid., pp. 37, 58.

30

Stewart, "Politics and the park," p. 142. See also Paul Boyer, *Urban masses and moral order in America, 1820–1920* (Cambridge, Mass.: Harvard University Press, 1978).

31

Rosenzweig and Blackmar, *The park and the people,* p. 55.

32

Ibid., p. 151.

33

See Stewart, "Politics and the park."

34

Political pressure to begin park construction also stemmed in part from the hope of creating thousands of park construction jobs in a time of high unemployment. See Rosenzweig and Blackmar, *The park and the people,* p. 58.

35

In 1857 the commission rejected the original plan of city engineer Egbert Viele, who continued to work on the project in bitter opposition to Olmsted and Vaux. Vaux had succeeded in convincing the commission that Viele's original proposal lacked any kind of overall aesthetic conception, and a competition was held to find an improved design. The choice of Olmsted and Vaux's design rested in part on their strong ties to the Republican park commissioners as well as their promotion of English naturalistic design traditions favored by Yankee elites in control of the State Legislature. See Rosenzweig and Blackmar, *The park and the people,* pp. 119–120, 130. On cultural and intellectual tensions in New York see also Thomas Bender, *New York intellect: a history of intellectual life in New York City, from 1750 to the beginnings of our own time* (Baltimore: Johns Hopkins University Press, 1981).

36

Rosenzweig and Blackmar, *The park and the people.* For alternative traditions in urban design see Chadwick, *The park and the town;* Aldo Rossi, *The architecture of the city,* trans. Diane Ghirardo and Joan Ockman (1966; Cambridge, Mass.: MIT Press, 1982); and Schaffer, ed., *Two centuries of American planning.*

37

In its scope and complexity the project rivaled the construction of the Croton Aqueduct and marked the emergence of landscape architecture as a new profession, just as the provision of water supply had led to a prominent role for civil engineers in urban government. By 1858, Central Park had become "the largest and most visible American experiment in the centralized corporate model for building public works," involving not only an unprecedented intervention in the private land market but also the development of new kinds of managerial and professional expertise. Olmsted would refer to park construction as an efficient "machine" in his later reminiscences of the project: a theme intensified from 1860 onward under Olmsted's successor Andrew Haswell Green (1820–1903), who introduced an even more dictatorial management regime under which workers' pay was routinely reduced in response to wider economic perturbations caused by wartime disruption of the cotton trade. See Rosenzweig and Blackmar, *The park and the people,* pp. 157–158, 180–205.

38

C. Rawolle and Ignaz A. Pilat, *Catalogue of plants gathered in August and September 1857 in the ground of Central Park* (New York: M. W. Siebert, 1857).

39

Perkins, *The Central Park,* pp. 12–13.

40

See Rosenzweig and Blackmar, *The park and the people,* pp. 65–83.

41

Zetzel, "The garden in the machine," p. 296.

42

Ibid., p. 321.

43

Kelly, "Art of the Olmsted landscape"; Judith K. Major, *To live in the New World: A. J. Downing and American landscape gardening* (Cambridge, Mass.: MIT Press, 1997).

44

See Frederick Law Olmsted, *Walks and talks of an American farmer in England* (1852; Ann Arbor: University of Michigan Press, 1967). For greater detail on the English picturesque, see Malcolm Andrews, *The search for the picturesque: landscape aesthetics and tourism in Britain, 1760–1800* (Stanford, Calif.: Stanford University Press, 1989); Ann Bermingham, *Landscape and ideology: the English rustic tradition, 1740–1860* (Berkeley: University of California Press, 1986); Stephen Copley and Peter Garside, eds., *The politics of the picturesque: literature, landscape and aesthetics since 1770* (Cambridge: Cambridge University Press, 1994); Ralph Dutton, *The English garden* (London: Batsford, 1937); John Dixon Hunt and Peter Willis, eds., *The genius of the place: the English landscape garden, 1620–1820* (1975; Cambridge, Mass.: MIT Press, 1988); Hunt, *Gardens and the picturesque;* John Dixon Hunt, ed., *The pastoral landscape* (Washington: National Gallery of Art, 1992); Sidney K. Robinson, *Inquiry into the picturesque* (Chicago: University of Chicago Press, 1991); Christopher Tunnard, *Gardens in the modern landscape* (London: Architectural Press, 1948); and David Watkin, *The English vision: the picturesque in architecture, landscape, and garden design* (New York: Harper and Row, 1982).

45

For a full account of the influence and significance of Repton's ideas on the history of landscape design see Stephen Daniels, *Humphry Repton: landscape gardening and the geography of Georgian England* (New Haven: Yale University Press, 1999).

46

See Louise C. Burnham and George W. Packard, *Central Park: a visit to one of the world's most treasured landscapes* (New York: Crescent Books, 1993); Stephen Rettig, "Influences across the water: Olmsted and England," in Kelly, Guillet, and Hern, eds., *Art of the Olmsted landscape,* pp. 79–85; and Louise Wyman, "The dialogue between society and ideological vision," *A+U: Architecture and Urbanism* 312 (1996): 122–127.

47

Robert Smithson, "Frederick Law Olmsted and the dialectical landscape" (1973), in *Robert Smithson: the collected writings,* ed. Jack Flam (Berkeley: University of California Press, 1995), p. 160

48

For Olmsted and the "genteel reformers" of nineteenth-century New York, Central Park was conceived as a visual emblem of taste and refinement and as a mark of civic maturity. From Burke, Hazlitt, and Reynolds onward we find a distinctive combination of political and aesthetic issues stemming from conceptions of a public vision that enabled individuals "to see beyond particular local contingencies and merely individual interest." See the critique of Thomas Crow developed by Deutsche in *Evictions,* pp. 306–307. The idea that the original conception of Central Park embodied a civic ideal is closely bound up with a powerful strand of Enlightenment aesthetics and political theory. John Barrell notes, for example, that women were denied full citizenship rights because "they were believed incapable of generalizing from particulars and therefore of exercising public vision" (cited in Deutsche, *Evictions,* p. 311). Furthermore, the conception of public space as the embodiment of reason draws on a distinctively Hegelian understanding of the role of the state in modern societies. See Dimendberg, "Henri Lefebvre on abstract space," p. 26.

49

The American romantic tradition is explored in Lawrence Buell, *The environmental imagination: Thoreau, nature writing, and the formation of American culture* (Cambridge, Mass.: Harvard University Press, 1995); Beth Lynne Leuck, *American writers and the picturesque tour: the search for national identity, 1790–1860* (New York: Garland, 1997); Roderick Nash, *Wilderness and the American mind,* 3d ed. (New Haven: Yale University Press, 1982), pp. 58–89; Blake Nevius, *Cooper's landscapes: an essay on the picturesque vision* (Berkeley: University of California Press, 1976); Barbara Novak, *Nature and culture: American landscape and painting, 1825–1875* (New York: Oxford University Press, 1980); Kenneth John Myers, "On the cultural construction of landscape experience: contact to 1830," in David C. Miller, ed., *American iconology: new approaches to nineteenth-century art and literature* (New Haven: Yale University Press, 1993); Sue Rainey, *Creating picturesque America: monument to the natural and cultural landscape* (Nashville: Vanderbilt University Press, 1994); Roger B. Stein, *John Ruskin and aesthetic thought in America, 1840–1900* (Cambridge, Mass.: Harvard University Press, 1967); Thomas Weiskel, *The romantic sublime: studies in the structure and psychology of transcendence* (Baltimore: Johns Hopkins University Press, 1976); Bryan Jay Wolf, *Romantic re-vision: culture and consciousness in nineteenth-century American painting and literature* (Chicago: University of Chicago Press, 1982).

50

Leo Marx, *The machine in the garden: technology and the pastoral ideal in America* (Oxford: Oxford University Press, 1964), p. 222.

51

Ibid., p. 139.

52

Cleaveland, "The Central Park," p. 691.

53

See James T. Flexner, *That wilder image: the painting of America's nature school from Thomas Cole to Winslow Homer* (Boston: Little, Brown, 1962); Kenneth W. Maddox, *In search of the picturesque: nineteenth-century images of industry along the Hudson River Valley* (Annandale-on-Hudson, N.Y.: Bard College, 1983); Jean Gardner McClintock, "Olmsted on the road: a view of Paradise," in Kelly, Guillet, and Hern, eds., *Art of the Olmsted landscape;* E. C. Parry III, *The art of Thomas Cole: ambition and imagination* (Newark: University of Delaware Press, 1988); W. H. Truettner and A. Wallach, eds., *Thomas Cole: landscape into history* (New Haven: Yale University Press, 1994).

54

The term "urban pastoral" is used here to denote a kind of intermediary aesthetic sensibility between rural nostalgia and the emergence of twentieth-century urban visions focused on new kinds of urban modernities facilitated by greater light, speed, and space. See John Archer, "Country and city in the American romantic suburb," *Journal of the Society of Architectural Historians* 42 (1983): 139–156; Michael H. Cowan, *City of the west: Emerson, America and the urban metaphor* (New Haven: Yale University Press, 1967); Marx, *The machine in the garden;* Charles A. Miller, *Jefferson and nature: an interpretation* (Baltimore: Johns Hopkins University Press, 1988); David E. Nye, *American technological sublime* (Cambridge, Mass.: MIT Press, 1994); Simon Pugh, ed., *Reading landscape: country, city, capital* (Manchester: Manchester University Press, 1990); Simon Schama, *Landscape and memory* (New York: Knopf, 1995); Peter J. Schmitt, *Back to nature: the Arcadian myth in urban America* (New York: Oxford University Press, 1969); and Raymond Williams, *The country and the city* (Oxford: Oxford University Press, 1973). Thomas Cole's *Oxbow* is a critical icon in the early nineteenth-century search for a distinctively American landscape aesthetic. Above all, the image depicts a real ambivalence toward material progress. See Stephen Daniels, *Fields of vision: landscape imagery and national identity in England and the United States* (Cambridge: Polity Press, 1993). On the significance of landscape for the evolution of American cultural identity, see also Novak, *Nature and culture;* and David Wyatt, *The fall into Eden: landscape and imagination in California* (Cambridge: Cambridge University Press, 1986).

55

Williams, *The country and the city,* p. 125.

56

See Thomas Bender, *Toward an urban vision: ideas and institutions in nineteenth-century America* (Baltimore: Johns Hopkins University Press, 1987); Burnham and Packard, *Central Park: a visit;* and Fein, *Frederick Law Olmsted and the American environmental tradition.*

57

Frederick Law Olmsted to Parke Godwin, 1 August 1858, cited in *The Papers of Frederick Law Olmsted,* vol. 3, *Creating Central Park* (1857–1861), ed. Charles E. Beveridge and David Schuyler (Baltimore: Johns Hopkins University Press, 1983), pp. 200–201.

58

Frederick Law Olmsted, *The cotton kingdom: a traveller's observations on cotton and slavery in the American slave states* (1861; New York: Knopf, 1953). See Fein, *Frederick Law Olmsted and the American environmental tradition.*

59

Albert Fein, for example, places Olmsted in the tradition of "stewardship" with respect to both nature and society associated with Thomas Jefferson, Abraham Lincoln, and Theodore Roosevelt; see Fein, "The Olmsted renaissance." Olmsted was a founding member of the American Social Science Association, whose charter emphasized "the responsibilities of the gifted and educated classes toward the weak, the witless, and the ignorant." Cited in Blodgett, "Frederick Law Olmsted," p. 875. When Olmsted addressed the association in 1870 at the Lowell Institute in Boston he noted how the park "exercises a distinctly harmonizing and refining influence upon the most unfortunate and lawless classes of the city." Cited in Melvin Kalfus, *Frederick Law Olmsted: the passion of a public artist* (New York: New York University Press, 1990), p. 278.

60

Frederick Law Olmsted, cited in Rosenzweig and Blackmar, *The park and the people,* p. 138.

61

Fitch, "Design and designer"; Broadus Mitchell, *Frederick Law Olmsted: a critic of the old South* (Baltimore: Johns Hopkins University Press, 1924). See also Iver Bernstein, *The New York City draft riots: their significance for American society and politics in the age of the Civil War* (Oxford: Oxford University Press, 1990). One must add that Mayor Fernando Wood, a critical political force behind the creation of Central Park, was an outspoken defender of slavery on the grounds of New York's economic reliance on the cotton trade. See Jerome Mushkat, *Fernando Wood: a political biography* (Kent, Ohio: Kent State University Press, 1990).

62

Quoted in Zetzel, "The garden in the machine," p. 311.

63

Frederick Law Olmsted, "The people's park at Birkenhead, near Liverpool," *Horticulturalist* 6 (1851): 224–228; Olmsted, *Walks and talks of an American farmer.* See also Hazel Conway, *People's parks: the design and development of Victorian parks in Britain* (Cambridge: Cambridge University Press, 1991); Rettig, "Influences across the water."

64

Perkins, *The Central Park,* p. 24.

65

Zetzel, "The garden in the machine," p. 323.

66

Cited in Alfred Kazin, *A writer's America: landscape in literature* (New York: Alfred A. Knopf, 1988), p. 171.

67

Some indication of the extent of controls on park use is suggested by the types of ordinance issued by the park commissioners. See Board of Commissioners of the Central Park (1865; held in the archives of the New-York Historical Society):

> § 15. No person shall be allowed to tell fortunes or play any game of chance at or with any table or instrument of gaming, nor do any obscene or indecent act whatever on the Central Park
>
> § 26. Ne [sic] person shall, without the consent of the Comptroller of the Park, play upon any musical instrument within the Central Park, nor shall any person take into, or carry or display in the Central Park, any flag, banner, target, or transparency.
>
> § 31. No person on foot shall go upon the grass, lawn, or turf of the Central Park, except when and where the word "common" is posted, indicating that persons are at liberty at that time and place to go on the grass.

68

Chadwick, *The park and the town.*

69

See Edwin G. Burrows and Mike Wallace, *Gotham: a history of New York City to 1898* (New York: Oxford University Press, 1999); Mona Domosh, "Those 'gorgeous incongruities': polite politics and public space on the streets of nineteenth-century New York City," *Annals of the Association of American Geographers* 88 (1998): 209–226.

70

Olmsted was writing here as architect-in-chief for the Central Park Commission's *Second Annual Report.* Cited in *The Papers of Frederick Law Olmsted,* vol. 3, p. 213. Even after Olmsted's departure, however, his supporters managed to defeat Tammany-backed legislation to build a racetrack in the park. Much of the political infighting surrounding the park also stemmed from the issue of who would control its employment patronage. See Blodgett, "Frederick Law Olmsted," pp. 882–883; *The Central Park race track law was repealed by public sentiment* (New York: Albert B. King, 1892); Rosenzweig and Blackmar, *The park and the people,* p. 154.

71

See Bender, *Toward an urban vision.*

72

Cited in Blodgett, "Frederick Law Olmsted," p. 882.

73

See Mona Domosh, *Invented cities: the creation of landscape in nineteenth-century New York and Boston* (New Haven: Yale University Press, 1996). After the Civil War there was a rapid development of new transportation and communication networks: the railroad, the telephone, and the telegram all played their part in a complex reordering of the relative significance and prosperity of different urban centers. While cities such as St. Louis saw their economic importance wane, New York assumed an increasingly dominant position within the evolving North American urban hierarchy (shortly after 1900, New York emerged as the largest port in the world). As the economic role of the city changed, the relative power of the city's political and economic elites shifted decisively. See Boyer, *Dreaming the rational city;* William Cronon, *Nature's metropolis: Chicago and the Great West* (New York: Norton, 1991); James Lemon, "Liberal dreams and nature's limits: great cities of North America since 1600," *Annals of the Association of American Geographers* 86 (1996): 745–866.

74

Blodgett, "Frederick Law Olmsted," p. 870.

75

Ibid., p. 888.

76

Cited in Kalfus, *Frederick Law Olmsted,* p. 274. See also Bender, *New York intellect.*

77

Boyer, *Dreaming the rational city,* p. 20. See also Patricia Burgess, "The expert's vision: the role of design in the historical development of city planning," *Journal of Architectural and Planning Research* 14 (1997): 91–106; W. H. Wilson, *The city beautiful movement* (Baltimore: Johns Hopkins University Press, 1989).

78

Terence Young, "Trees, the park and moral order: the significance of Golden Gate Park's first plantings," *Journal of Garden History* 14 (1994): 158–170.

79

Smithson, "Frederick Law Olmsted and the dialectical landscape," p. 170.

80

Douglas Martin, "Central Park entrances in a return to the past," *New York Times,* December 3, 1999.

81

The Central Park Conservancy was created in 1980 as a partnership between the city and private sources of funding in the aftermath of the city's fiscal crisis in the 1970s. The shift in emphasis toward the restoration of the original greensward ideal can to be traced to 1965 and the designation of Central Park as the city's first Scenic Landmark and also a National Landmark listed under the National Register of Historic Places. See James Bradley, "Faded glory: is privatization going to solve New York's park problems," *City Limits* (August/September 1994), pp. 30–32; Cindi Katz, "Whose nature, whose culture? Private production of space and the 'preservation' of nature," in Bruce Braun and Noel Castree, eds., *Remaking reality: nature at the millennium* (London: Routledge, 1998), pp. 46–63.

82

Blaine Harden, "Neighbors give Central Park a wealthy glow," *New York Times,* 22 November 1999.

83

Herbert Muschamp, "From stereotypes of urban decay to signs of life," *New York Times,* 3 April 1994.

84

Harden, "Neighbors give Central Park a wealthy glow."

85

"Hundreds gather to protest city's auction of garden lots," *New York Times,* 11 April 1999; "Giuliani seeks deal to sell 63 gardens to land group and end suits," *New York Times,* 12 May 1999.

86

Andrew Jacobs, "Gentrification led to the unrest at Tompkins Square 10 years ago. Did the protesters win that battle but lose the war?" *New York Times,* 9 August 1998; Neil Smith, *The new urban frontier: gentrification and the revanchist city* (London: Routledge, 1996).

87

Don Mitchell points out that "the public sphere in the American past was anything but inclusive." See Mitchell, "The end of public space," p. 121. On the history of cities and the public sphere see also Deutsche, *Evictions;* Sara M. Evans and Harry C. Boyte, eds., *Free spaces: the sources of democratic change in America* (Chicago: University of Chicago Press, 1992); Nancy Fraser, "Rethinking the public sphere," in Craig Calhoun, Peter E. Hohendahl, and Benjamin Lee, eds., *Habermas and the public sphere* (Cambridge, Mass.: MIT Press, 1992); Phil Howell, "Public space and the public sphere: political theory and the historical geography of modernity," *Environment and Planning D: Society and Space* 11 (1993): 303–322; Claude Lefort, *The political forms of modern society: bureaucracy, democracy, totalitarianism* (Cambridge, Mass.: MIT Press, 1986); Oskar Negt and Alexander Kluge, *Public sphere and experience: toward an analysis of the bourgeois and proletarian public sphere,* trans. P. Labanyi, J. O. Daniel, and A. Oksiloff (Minneapolis: University of Minnesota Press, 1993); Bruce Robbins, ed., *The phantom public sphere* (Minneapolis: University of Minnesota Press, 1993); Richard Sennett, *The fall of public man: on the social psychology of capitalism* (1976; New York: W. W. Norton, 1992); Richard Sennett, *The conscience of the eye: the design and social life of cities* (New York: Alfred A. Knopf, 1990); Richard Sennett, *Flesh and stone: the body and the city in western civilization* (London: Faber and Faber, 1994); Michael Sorkin, ed., *Variations on a theme park: the new American city and the end of public space* (New York: Hill and Wang, 1992).

88

Mike Davis, *City of quartz: excavating the future in Los Angeles* (London: Verso, 1990), p. 231.

89

Marshall Berman, "Take it to the streets: conflict and community in public space," *Dissent* 33 (1986): 476–485.

90

Bronx Republican leader Paul Fino, cited in Rosenzweig and Blackmar, *The park and the people,* p. 495.

91

Dominic A. Pacyga, "Central Park and the public sphere," *Journal of Urban History* 22 (1996): 649–655.

92

See Edward S. Casey, "The production of space *or* the heterogeneity of place: a commentary on Edward Dimendberg and Neil Smith," in Light and Smith, eds., *Philosophy and geography II,* pp. 50–81; Smith, "Antinomies of space and nature in Henri Lefebvre's *The production of space.*"

93

Koolhaas, *Delirious New York,* p. 21.

94

Peter Calthorpe, *The next American metropolis: ecology, community and the American dream* (New York: Princeton Architectural Press, 1993), p. 25.

95

Charles Jencks, *Heteropolis* (London: Academy Editions, 1993).

96

Reyner Banham, *Los Angeles: the architecture of four ecologies* (Harmondsworth: Penguin, 1971).

3

TECHNOLOGICAL MODERNISM AND THE URBAN PARKWAY

1

J. G. Ballard, *Crash* (1973; London: Vintage, 1995), p. 76.

2

Hamilton Wright quoted in the *American City* 36 (1927): 801. Wright's description was of a drawing by Harvey Wiley Corbett that depicted the Manhattan of the future.

3

American discourses of art and design began to loosen "the grip of a rigid academic traditionalism" exemplified by the Chicago World's Columbian Exposition of 1893. See Garrett Eckbo, *Public landscape: six essays on government and environmental design in the San Francisco Bay area* (Berkeley: University of California, Berkeley, Institute of Governmental Studies, 1978), p. 4.

4

Klaus-Jürgen Sembach, *Into the thirties: style and design, 1927–1934,* trans. Judith Filson (London: Thames and Hudson, 1972).

5

See, for example, Marshall Berman, *All that is solid melts into air: the experience of modernity* (London: Verso, 1982); Sigfried Giedion, *Space, time and architecture* (Cambridge, Mass.: Harvard University Press, 1941); Terry Smith, *Making the modern: industry, art, and design in America* (Chicago: University of Chicago Press, 1993); Richard Guy Wilson, Dianne H. Pilgrim, and Dickran Tashjian, *The machine age in America, 1918–1941* (New York: Brooklyn Museum and Harry N. Abrams, 1986).

6

Mikael Hård and Andrew Jamison, "Conceptual framework: technology debates as appropriation process," in Mikael Hård and Andrew Jamison, eds., *The intellectual appropriation of technology: discourses on modernity, 1900–1939* (Cambridge, Mass.: MIT Press, 1998), pp. 1–15; Peter Wagner, "Sociological reflections: the technology question during the first crisis of modernity," in Hård and Jamison, *The intellectual appropriation of technology*, pp. 225–252.

7

Peter S. Reed, "Enlisting modernism," in Donald Albrecht, ed., *World War II and the American dream: how wartime building changed a nation* (Cambridge, Mass.: MIT Press; Washington, D.C: National Building Museum, 1995), pp. 2–41; Gwendolyn Wright, "Inventions and interventions: American urban design in the twentieth century," in Russell Ferguson, ed., *Urban revisions: current projects for the public realm* (Cambridge, Mass.: MIT Press; Los Angeles: Museum of Contemporary Art, 1994), pp. 26–37.

8

On the legacy of modernist landscape design in America see Catherine Howett, "Modernism and American landscape architecture," in Marc Treib, ed., *Modern landscape architecture: a critical review* (Cambridge, Mass.: MIT Press, 1993), pp. 2–35; Reuben M. Rainey, "'Organic form in the humanized landscape': Garrett Eckbo's *Landscape for living*," in Treib, ed., *Modern landscape architecture*, pp. 180–205; Marc Treib, "Axioms for a modern landscape architecture," in Treib, ed., *Modern landscape architecture*, pp. 36–67; Peter Walker and Melanie Simo, *Invisible gardens: the search for modernism in the American landscape* (Cambridge, Mass.: MIT Press, 1994); Alexander Wilson, *The culture of nature: North American landscape from Disney to Exxon Valdez* (Oxford and Cambridge, Mass.: Basil Blackwell, 1992). Other key figures in the development of modernist landscape design include Fletcher Steele, Thomas Church, Christopher Tunnard, James Rose, and Dan Kiley. Much of the new modernist architecture and landscape design was based in California, with significant input from European émigrés from the 1930s onward associated with the Bauhaus and other elements of the cultural avant-garde.

9

On the pivotal role of New York City in the creation of the modern urban landscape see Robert A. M. Stern, Gregory Gilmartin, and Thomas Mellins, *New York, 1930: architecture and urbanism between the two world wars* (New York: Rizzoli, 1987). On the history of cars and the transformation of urban space see, for example, Scott L. Bottles, *Los Angeles and the automobile: the making of the modern city* (Berkeley: University of California Press, 1987); Keller Easterling, *Organization space: landscape, houses, and highways in America* (Cambridge, Mass.: MIT Press, 1999); James J. Flink, *The car culture* (Cambridge, Mass.: MIT Press, 1975); James J. Flink, *The automobile age* (Cambridge, Mass.: MIT Press, 1988); Jan Jennings, ed., *Roadside America: the automobile in design and culture* (Ames: Iowa State University Press, 1990); Maxwell G. Lay, *Ways of the world: a history of the world's roads and of the vehicles that used them* (New Brunswick, N.J.: Rutgers University Press, 1992); Clay McShane, *Down the asphalt path: the automobile and the American city* (New York: Columbia University Press, 1994); C. W. Pursell, ed., *Readings in technology and American life* (London: Open University Press, 1969); W. Sachs, *For love of the automobile* (Berkeley: University of California Press, 1992).

10

Berman, *All that is solid melts into air,* p. 294.

11

Blair A. Ruble, "Failures of centralized metropolitanism: inter-war Moscow and New York," *Planning Perspectives* 9 (1994): 366.

12

Cited in Paul Ramon Pescatello, "Westway: the road from New Deal to new politics" (PhD dissertation, Cornell University, 1986), p. 119.

13

Lewis Mumford, cited in Garrett Eckbo, Daniel U. Kiley, and James C. Rose, "Landscape design for the urban environment," *Architectural Record* (May 1939), pp. 70–81.

14

The engineer Spencer Miller, Jr., notes that the State of New Jersey took a lead in highway legislation by passing in 1889 a general county road law permitting individual counties to issue bonds for road construction. The growing significance of state intervention in American road building is indicated by the Federal Aid Road Act of 1916 and the gradual evolution of the Office of Road Inquiry of 1893 into the Office of Public Roads in 1905. Spencer Miller, Jr., "History of the modern highway in the United States," in Jean Labatut and Wheaton A. Lane, eds., *Highways in our national life* (Princeton: Princeton University Press, 1950), pp. 88–119. See also

Gilmore D. Clarke, "The parkway idea," in W. A. Bugge and W. Brewster Snow, eds., *The highway and the landscape* (New Brunswick, N.J.: Rutgers University Press, 1959), pp. 33–55; Christian Zapatka, "The American parkways: origins and evolution of the park road," *Lotus International* 56 (1988): 96–128.

15

City of Brooklyn, *Fourteenth Annual Report of the Brooklyn Park Commissioners* (Brooklyn: Printed for the Commissioner, 1874). The completion of the Brooklyn parkways exposed tensions between land speculation and agricultural interests as the urban fringe extended into predominantly rural areas in close proximity to Manhattan Island. See Marc Linder and Lawrence S. Zacharias, *Of cabbages and Kings County: agriculture and the formation of modern Brooklyn* (Iowa City: University of Iowa Press, 1999). Maxwell G. Lay points out that the term "parkway" originated in Williamsburg, Virginia, in 1699 and was applied to roads with "wide, grassy central medians." Lay, *Ways of the world,* p. 314.

16

Zapatka, "The American parkways."

17

Jay Downer, "Reclaiming a polluted river," *American City* 22 (1920): 15. See also Jay Downer, "The Bronx River Parkway," *Municipal Engineers Journal* 2 (1916): 1–20; Leslie G. Holleran, "Construction plans developed for the Bronx River Parkway Reservation," *Engineering News Record* (14 November 1918), pp. 899–903; Hermann W. Merkel, "The New York idea of a Zoölogical park," *American City* 9 (1913): 298–302; New York State Bronx Parkway Commission, *Report of the Bronx Parkway Commission* (Albany: J. B. Lyon, 1909); Rebecca B. Rankin, "Bronx Parkway Commission," manuscript prepared for the Municipal Reference Library, New York, 1919; George E. Spargo, "Parks and recreational facilities of the city of New York," *Municipal Engineers Journal* 25 (1938): 67–85; Bronx Board of Trade, *Parks and parkways in the Borough of the Bronx, New York City* (1914); Marilyn E. Weingold, *Pioneering in parks and parkways: Westchester County, New York, 1895–1945* (Chicago: Public Works Historical Society, 1980).

18

New York State Bronx Parkway Commission, *Final Report of the Bronx Parkway Commission* (1925), p. 24.

19

Hermann W. Merkel, cited in ibid., p. 66.

20

Gilmore D. Clarke, "Modern motor ways," *Architectural Record* 74 (December 1933): 434.

21

Howett, "Modernism and American landscape architecture."

22

On the contrived character of naturalistic design see Garrett Eckbo, Daniel U. Kiley, and James C. Rose, "Landscape design for the rural environment," *Architectural Record* (August 1939), pp. 70–71.

23

Clarke, "Modern motor ways," 435.

24

Gert Gröning and Joachim Wolschke-Bulmahn, "Politics, planning and the protection of nature: political abuse of early ecological ideas in Germany, 1933–1945," *Planning Perspectives* 2 (1987): 128–129; Gert Gröning, "The feeling for landscape: a German example," *Landscape Research* 17 (1992): 108–115; Thomas Lekan, "Regionalism and the politics of landscape preservation in the Third Reich," *Environmental History* 4 (1999): 384–404; William H. Rollins, "Whose landscape? Technology, fascism, and environmentalism on the National Socialist autobahn," *Annals of the Association of American Geographers* 85 (1995): 494–520.

25

Zapatka, "The American parkways," p. 115.

26

See Clarke, "The parkway idea." See also Thomas J. Campanella, "American curves: Gilmore D. Clarke and the modern civic landscape," *Harvard Design Magazine* (Summer 1997), p. 40.

27

Eckbo, *Public landscape,* p. 89.

28

Giedion, *Space, time and architecture,* pp. 550, 552.

29

Zapatka, "The American parkways," p. 97. See also Sylvia Crowe, *The landscape of roads* (London: Architectural Press, 1960); Clarke, "The parkway idea"; Labatut and Lane, eds., *Highways in our national life;* Lay, *Ways of the world.* The Arroyo Seco Parkway between Pasadena and downtown Los Angeles was in effect the beginning of the LA freeway system. See David Brodsly, *L.A. freeway: an appreciative essay* (Berkeley: University of California Press, 1981).

30

Giedion, *Space, time and architecture,* p. 554.

31

William S. Chapin, "The expressway and parkway program for metropolitan New York," *Municipal Engineers Journal* (1947), pp. 21–26.

32

M. Christine Boyer, *Dreaming the rational city: the myth of American city planning* (Cambridge, Mass.: MIT Press, 1983), p. 183; Ruble, "Failures of centralized metropolitanism," p. 368.

33

Centre de Cultura Contemporània de Barcelona, *Visions urbanes. Europa, 1870–1993. La ciutat de l'artista. La ciutat de l'arquitecte* (Barcelona: Centre de Cultura Contemporània de Barcelona, 1994); Mikael Hård, "German regulation: the integration of modern technology into national culture," in Hård and Jamison, eds., *The intellectual appropriation of technology,* pp. 33–67; Jeffrey Herf, *Reactionary modernism: technology, culture and politics in Weimar and the Third Reich* (Cambridge: Cambridge University Press, 1984).

34

Edward Dimendberg, "The will to motorization: cinema, highways and modernity," in Jeremy Millar and Michiel Schwarz, eds., *Speed: visions of an accelerated age* (London: Whitechapel Art Gallery and the Photographers' Gallery, 1998), p. 60.

35

Centre de Cultura Contemporània de Barcelona, *Visions urbanes.* The Long Island parkways employed several examples of the cloverleaf intersection designed by Eugène Hénard. See Wilson, Pilgrim, and Tashjian, *The machine age in America,* p. 97; Peter M. Wolf, *Eugène Hénard and the beginning of urbanism in Paris, 1900–1914* (The Hague: International Federation for Housing and Planning; Paris: Centre de Recherche d'Urbanisme, 1968).

36

Luciano Caramel and Alberto Longatti, *Antonio Sant'Elia* (Como: Villa Comunale dell'Olmo, 1962).

37

Harvey Wiley Corbett, "Different levels for foot, wheel and rail," *American City* 31 (1924): 2–6.

38

Dr. John A. Harriss, quoted in the *American City* 36 (1927): 803.

39

Le Corbusier, *The radiant city,* trans. Pamela Knight, Eleanor Levieux, and Derek Coltman (1935; London: Faber and Faber, 1967). See also Le Corbusier, *The city of tomorrow and its plan-*

ning (1925; New York: Payson & Clark, 1929), and Le Corbusier, *Towards a new architecture* (1923; New York: Warren and Putnam, 1927). See also William J. R. Curtis, *Le Corbusier: ideas and forms* (London: Phaidon, 1986).

40

Dimendberg, "The will to motorization."

41

Norman Bel Geddes, *Magic motorways* (New York: Random House, 1940), p. 283.

42

Christian Zapatka, "In progress's own image: the New York that Robert Moses built," *Lotus International* 89 (1996): 102–131.

43

Ruble, "Failures of centralized metropolitanism," p. 361.

44

Spargo, "Parks and recreational facilities," p. 84.

45

Berman, *All that is solid melts into air,* p. 299.

46

Robert Caro, "The city-shaper," *New Yorker* (January 5, 1998).

47

Berman, *All that is solid melts into air,* p. 296.

48

Lewis Mumford, "The sky line: bridges and beaches," cited in Stern, Gilmartin, and Mellins, *New York, 1930,* p. 722.

49

See James S. Russell, "Reconsidering Robert Moses: power vs. paralysis," *Architectural Record* 177 (1989): 49–50.

50

Leonard Wallock, "The myth of the master builder: Robert Moses, New York and the dynamics of metropolitan development since World War II," *Journal of Urban History* 17 (1991): 339–362; New York City Planning Commission, *Adoption of a system of express highways, parkways and major streets and of a city-wide map thereof as part of the master plan* (1941); New York Metropolitan Conference on Parks, *Program for extension of parks and parkways in the metropolitan region* (1930).

51

Harry W. Levy, "Manhattan—past, present, future," *Municipal Engineers Journal* 11 (1925): 125–150; George T. Hammond, "City planning and the engineer," *Municipal Engineers Journal* 2 (1916): 332–369; William J. Ronan, "Scientific management comes to New York City," *Municipal Engineers Journal* 42 (1956): 163–171; D. L. White, *City Manager* (Chicago: University of Chicago Press, 1927).

52

Boyer, *Dreaming the rational city,* p. 184.

53

Zapatka, "The American parkways," p. 105.

54

Although the financing of the Triborough Bridge began in 1929, it was suspended until the federal government launched the first public works program through the Reconstruction Finance Corporation and the appointment of the State Emergency Public Works Commission by Governor Herbert Lehman in 1933. After it had been established that the city was not in a position to finance the undertaking with its own bonds or corporate stock, it was agreed that an authority should be set up by state law authorized to issue its own bonds which should not involve the credit of the city. This new public authority was the Triborough Bridge Authority, later consolidated as the Triborough Bridge and Tunnel Authority after the merger of the Triborough Bridge Authority (created 1933) and the New York City Tunnel Authority (created 1936). See David C. Perry, "Robert Moses. The public authority and the project at hand: bridge-building in New York in the 1930s," *Intersight* 2 (1993): 84–93; Triborough Bridge Authority, *The Triborough Bridge: a modern metropolitan traffic artery* (1936); Zapatka, "In progress's own image."

55

Robert Moses, "Statement with reference to the proposed, 1938 parkway construction program in the metropolitan area" (February 10, 1938, New York City Department of Parks). Moses described how "it has long been an ambition of mine to weave together the loose strands and frayed edges of New York's metropolitan arterial tapestry"; see Triborough Bridge Authority, *The Triborough Bridge Authority: fifth anniversary* (1941), p. 6. For details on design and construction see Triborough Bridge Authority, *Brooklyn Battery Bridge* (1939); Triborough Bridge Authority, *Improvement of Eastern Boulevard, the Bronx* (1940).

56

Giedion, *Space, time and architecture,* p. 557.

57

Robert A. Caro, *Power broker: Robert Moses and the fall of New York* (New York: Vintage, 1974); Joann P. Krieg, *Robert Moses: single-minded genius* (Interlaken, N.Y.: Heart of the Lakes Publishing, 1989). For a neo-Marxian perspective see also the interesting essay by Fredric Jameson, "The brick and the balloon: architecture, idealism and land speculation," in *The cultural turn: selected writings on the postmodern, 1983–1998* (London: Verso, 1998), pp. 162–189.

58

Wallock, "The myth of the master builder."

59

Ruble, "Failures of centralized metropolitanism." Other factors involved in this transformation were the social and spatial dimensions of the war economy in the 1940s, which underpinned new patterns of regional industrial development. See Greg Hise, "The airplane and the garden city: regional transformations during World War II," in Albrecht, ed., *World War II and the American dream,* pp. 144–183.

60

Brodsly, *L.A. freeway,* p. 135.

61

US Congress, *Interregional Highways,* House Document 379, 78th Congress, 2d Session, Washington, cited in Gilmore D. Clarke, "The design of motorways," in Labatut and Lane, eds., *Highways in our national life,* p. 308.

62

Robert Moses, "The changing city," *Architectural Forum* 72 (March 1940): 143.

63

Robert Moses, "Mr. Moses dissects the 'long-haired planners': the Park Commissioner prefers common sense to their revolutionary theories," *New York Times Magazine* (25 June 1944).

64

Zapatka, "In progress's own image," p. 129.

65

See Robert Fitch, *The assassination of New York* (London: Verso, 1993), and Joel Schwartz, *The New York approach: Robert Moses, urban liberals, and redevelopment of the inner city* (Columbus: Ohio State University Press, 1993).

66

See Kenneth T. Jackson, "The capital of capitalism: the New York metropolitan region, 1890–1940," in Anthony Sutcliffe, ed., *Metropolis, 1890–1940* (Chicago: University of Chicago

Press, 1984); Cleveland Rogers and Rebecca C. Rankin, *New York: the world's capital city* (New York: Harper & Brothers, 1948); David Ward and Oliver Zunz, eds., *Landscapes of modernity: essays on New York City, 1900–1940* (New York: Russell Sage Foundation, 1992). By the end of World War II the United States accounted for over fifty percent of the world's industrial production and New York City enjoyed a preeminent global role symbolized by the completion of the United Nations headquarters complex in 1950. See Michael Sorkin, "War is swell," in Albrecht, ed., *World War II and the American dream,* pp. 231–251.

67

Zapatka, "In progress's own image," p. 11.

68

Langdon Winner, "Do artifacts have politics?" (1977), reprinted in Donald MacKenzie and Judy Wajcman, eds., *The social shaping of technology: how the refrigerator got its hum* (Milton Keynes: Open University Press, 1985), pp. 26–38. On the wider dynamics of racial segregation in American cities see the useful review by M. Patricia Fernández-Kelly, "Migration, race, and ethnicity in the design of the American city," in Ferguson, ed., *Urban revisions,* pp. 16–25.

69

Berman, *All that is solid melts into air,* p. 299; see also Robert Fishman, *Bourgeois utopias: the rise and fall of suburbia* (New York: Basic Books, 1987).

70

Zapatka, "In progress's own image," p. 124.

71

Rexford G. Tugwell, "San Francisco as seen from New York," *National Conference on Planning* (proceedings of the conference held at San Francisco, 8–11 July 1940), pp. 182–188. On the ugliness of the American landscape see also Paul B. Sears, "Science and the new landscape," *Harper's Magazine* (July 1939), p. 207.

72

Lewis Mumford, "The highway and the city," in *The urban prospect* (1958; New York: Harcourt, Brace & World, 1968), p. 95.

73

Robert Moses, *Public works: a dangerous trade* (New York: McGraw-Hill, 1970).

74

Giedion, *Space, time and architecture,* p. 555.

75

Boyer, *Dreaming the rational city,* p. 280.

76

Pescatello, "Westway," p. 122. In spite of the disruption, Moses argued that "we have acted with humane understanding and firmness"; cited in Triborough Bridge and Tunnel Authority, *Cross Bronx Expressway, Alexander Hamilton Bridge and George Washington Bridge Bus Station* (17 January 1963), p. 4. Yet Moses acknowledged that "this is just about the end of the opportunity to do anything like a good arterial job. . . . We now face enormous opposition on the part of the public generally and quite a section of the press." Triborough Bridge and Tunnel Authority, *Throgs Neck Bridge* (22 October 1957), p. 8.

77

Robert Moses, *Public works and beauty* (New York: Triborough Bridge and Tunnel Authority, 1966), p. 1.

78

Ibid., p. 2.

79

"Wagner orders building of Manhattan expressway," *New York Times,* 26 May 1965; "Expressway plan faces new delay: mayor may put off approval of Lower Manhattan road because of Lindsay race," *New York Times,* 14 May 1965; "Mayor is revising expressway plan, " *New York Times,* 29 October 1965; Jacobs, *The death and life of great American cities,* cited on the website www.nycroads.com; Helen Leavitt, *Superhighway—super hoax* (Garden City, N.Y.: Doubleday, 1970); "Lower Manhattan road killed under state plan," *New York Times,* 25 March 1971.

80

Lawrence Halprin, *Cities* (Cambridge, Mass.: MIT Press, 1963), p. 201.

81

Pescatello, "Westway," p. 3. See also "Battle of the Westway: bitter 10-year saga of a vision on hold," *New York Times,* 4 June 1984; Jose Gomez-Ibanez and Marc Roberts, *West Side Highway proposal* (Case Studies in Public Policy and Management, Harvard University, Kennedy School of Government, 1989).

82

Letter to the editor of the *New York Times* from William C. Finneran, Jr., general manager of the General Contractors Association of New York Inc. (dated 16 January 1976). See also letter from New York State Governor Hugh L. Carey to Secretary of Transportation William T. Coleman (29 January 1976) (New York Municipal Archives); Jon Ciner, "Westway is underway but is it a fait accompli?," *Villager* (6 April 1978); "Koch, an opponent of Westway, now telling aides it is inevitable," *New York Times,* 3 April 1978.

83

Roberta Brandes Gratz, "How Westway will destroy New York: an interview with Jane Jacobs," *New York* (6 February 1978). See also Tom Wicker, "How not to manage a city," *New York Times,* 14 September 1975; Carol Creitzer, "The New York strangler: a network of roads," *Village Voice* (22 April 1968); Leavitt, *Superhighway—super hoax.*

84

Judith Bender, "Westway: the final, fatal turns," *New York Newsday,* 26 September 1985.

85

Sam Roberts, "The legacy of Westway: lessons from its demise," *New York Times,* 7 October 1985.

86

Ibid.

87

Carter Wiseman, "Where Westway went: a case study in changing urban priorities," *Architectural Record* (February 1986), pp. 81–83.

88

Grace Anderson, "Big park for the big apple," *Architectural Record* (1985), pp. 124–131.

89

Lewis Mumford, cited in Wilson, *The culture of nature,* p. 264. See also the moralistic indictment of postwar urbanism contained in Lewis Mumford, *The urban prospect* (New York: Harcourt, Brace & World, 1968), p. 238.

90

Robert Moses, *Tomorrow's cars and roads* (New York: Triborough Bridge Authority), p. 4 (reprinted from *Liberty Magazine,* 24 January 1947).

91

Howett, "Modernism and American landscape architecture."

92

Eckbo, *Public landscape,* p. 5. On the absorbtion of American modernism into the International Style see Vincent Scully, Jr., *Modern architecture* (New York: George Braziller, 1982).

93

John Dixon Hunt, "The dialogue of modern landscape architecture with its past," in Treib, ed., *Modern landscape architecture,* pp. 134–143.

94

Phoebe Cutler, *The public landscape of the New Deal* (New Haven: Yale University Press, 1985), p. 51.

95

The Gowanus Parkway did have its supporters. Elizabeth Mock, for example, wrote: "Proud symbol of a new age, the highway cuts above its dreary surroundings, its long slim legs withdrawn from the chaos." Elizabeth B. Mock, *The architecture of bridges* (New York: Museum of Modern Art, 1949), p. 67.

96

See, for example, Kenneth Frampton, "Moses and megalopolis: res privata and res publica," paper presented to the conference on Robert Moses held at Columbia University, New York, 10–11 February 1989.

97

Robert Moses, "What happened to Haussmann," *Architectural Forum* (July 1942), p. 57. See also Mark Kenneth Abbott, "The master plan: the life and death of an idea" (PhD dissertation, Purdue University, 1985); Boyer, *Dreaming the rational city*.

98

Berman, *All that is solid melts into air,* p. 328. See also Vincent Scully, *Architecture: the natural and the man-made* (New York: St. Martin's Press, 1991); Stuart Wrede and William Howard Adams, eds., *Denatured visions: landscape and culture in the twentieth century* (New York: Museum of Modern Art, 1991).

4

BETWEEN BORINQUEN AND THE *BARRIO*

1

David Perez, cited in Michael Abramson/Young Lords Party, *Palante: Young Lords Party* (New York: McGraw-Hill, 1971), p. 9.

2

Pedro Pietri, "Puerto Rican obituary," in ibid., p. 17.

3

¡Palante, siempre Palante! The Young Lords, director: Iris Morales (New York: Latino Education Network Service, 1996). My first encounter with the legacy of the Young Lords was derived from an interview with Analia Penchaszadeh of the New York City Environmental Justice Alliance held on 4 April 1995.

4

Manuel Castells, *The city and the grassroots: a cross-cultural theory of urban social movements* (Berkeley: University of California Press, 1983). Castells describes how "the wave of riots in the black ghettos started in Harlem and Bedford-Stuyvesant, New York, in 1964, and represented the most direct challenge ever posed to the American social order, an order historically based upon racial discrimination and ethnic fragmentation among the lower classes" (p. 50). Between 1964 and 1968 there were over 300 riots involving hundreds of thousands of people in 257 cities, leading to 52,000 arrests, 8,000 injured, and at least 220 killed. In 1968, in the wake of Martin Luther King's assassination, there were within one month 202 violent incidents in 172 cities with 27,000 arrests, 3,500 injured, and 43 deaths.

5

This theme of the paradoxical legitimation of social change in the 1960s is explored by M. Christine Boyer, *Dreaming the rational city: the myth of American city planning* (Cambridge, Mass.: MIT Press, 1983).

6

See, for example, Luis Aponte-Parés, "What's yellow and white and has land all around it? Appropriating place in the Puerto Rican *barrios*," *Centro: Journal of the Centro de Estudios Puertorriqueños* 7, no. 1 (1995): 9–20; Castells, *The city and the grassroots*, pp. 106–137; José E. Cruz, "A decade of change: Puerto Rican politics in Hartford, Connecticut, 1969–1979," *Journal of American Ethnic History* (Spring 1997), pp. 45–80; Mike Davis, "Magical urbanism: Latinos reinvent the US big city," *New Left Review* 234 (1999): 3–43; Marjorie Heins, *Strictly ghetto property: the story of "Los Siete de la Raza"* (Berkeley, Calif.: Ramparts Press, 1972); Joan Moore and Raquel Pinderhughes, eds., *In the barrios: Latinos and the underclass debate* (New York: Russell Sage Foundation, 1993); and Joseph A. Rodríguez, "Ethnicity and the horizontal city: Mexican Americans and the Chicano movement in San Jose, California," *Journal of Urban History* 21 (1995): 597–621.

7

See Virginia Sánchez Korrol, *From colonia to community: the history of Puerto Ricans in New York City* (Berkeley: University of California Press, 1994); Kal Wagenheim, *A survey of the Puerto Ricans on the US mainland in the 1970s* (New York: Praeger, 1975).

8

Charles H. Rector, *The story of beautiful Porto Rico: a graphic description of the garden spot of the world* (Chicago: Laird and Lee, 1898), p. 183.

9

See Edna Acosta-Belén and Carlos E. Santiago, "Merging borders: the remapping of America," *Latino Review of Books* (Spring 1995), pp. 2–12; Angelo Falcon, "A history of Puerto Rican politics in New York City: 1860s to 1945," in James Jennings and Monte Rivera, eds., *Puerto Rican politics in urban America* (Westport, Conn.: Greenwood Press, 1984), pp. 15–42; Sánchez Korrol, *From colonia to community;* and César Andreu Iglesias, ed., *Memorias de Bernardo Vega: contribución a la historia de la communidad puertorriqueña en Nueva York* (Río Pedras, Puerto Rico: Ediciones Huracán, 1984). For a general overview of the main social and physical features of the islands under Spanish occupation see, for example, Fray Iñigo Abbad y Lasierra, *Historia geografica, civil y natural de la isla de San Juan Bautista de Puerto-Rico* (Puerto Rico: Libreria de Acosta, 1866).

10

Robert T. Hill, *Cuba and Porto Rico with the other islands of the West Indies* (New York: Century, 1899), p. 145.

11

William Dinwiddie, *Puerto Rico: its conditions and possibilities* (New York: Harper, 1899), p. 9.

12

Leonard Wood, William H. Taft, Charles H. Allen, Perfecto Lacoste, and M. E. Beall, *Opportunities in the colonies and Cuba* (New York: Lewis, Scribner, 1902), p. 287. Other examples of this kind of "commercial geography" include Frederick A. Ober, *Puerto Rico and its resources* (New York: A. Appleton, 1899), and Thomas J. Vivian and Ruel P. Smith, *Everything about our new possessions: being a handy book on Cuba, Porto Rico, Hawaii and the Philippines* (New York: R. F. Fenno, 1899). For recent analysis of this literature see Lillian Guerra, "The promise and disillusion of Americanization: surveying the socioeconomic terrain of early twentieth-century Puerto Rico," *Centro: Journal of the Centro de Estudios Puertorriqueños* 11, no. 1 (1998): 9–32; Félix V. Matos Rodríguez, "Their islands and our people: US writing about Puerto Rico, 1898–1920," *Centro: Journal of the Centro de Estudios Puertorriqueños* 11, no. 1 (1998): 33–50.

13

Congressional Record, 57th Congress, April 1900, cited in "Puerto Rico: 'island paradise' of US imperialism," *Palante* 1 (1969): 18. See also David Healy, *Drive to hegemony: the United States in the Caribbean, 1898–1917* (Madison: University of Wisconsin Press, 1988); Gordon Lewis, *Notes on the Puerto Rican revolution: an essay on American dominance and Caribbean resistance* (New York: Monthly Review Press, 1974).

14

James L. Dietz, *Economic history of Puerto Rico* (Princeton, N.J.: Princeton University Press, 1986); Ronald Fernandez, *The disenchanted island: Puerto Rico and the United States,* 2d ed. (Westport, Conn.: Praeger, 1996); History Task Force, Center for Puerto Rican Studies, *Labor migration under capitalism: the Puerto Rican experience* (New York: Monthly Review Press, 1979); Déborah Berman Santana, "Geographers, colonialism, and development strategies: the case of Puerto Rico," *Urban Geography* 17 (1996): 456–474.

15

Rexford G. Tugwell, *The stricken land: the story of Puerto Rico* (1946; New York: Greenwood Press, 1977), p. 34.

16

Immigration was further encouraged under the Jones Act of 1917, which granted Puerto Ricans US citizenship in order to enlist conscripts and secure additional labor for the war industries. At the same time, continuing migration from southern and eastern Europe was restricted by the Quota Law of 1924 and the National Origins Act of 1929. During the interwar era Puerto Rican employment in New York became concentrated in the cigar industry, with settlement clustered around the Brooklyn Navy Yard, Red Hook, and especially the emerging *barrio* of East Harlem. These areas rapidly emerged as the focal point for Puerto Rican social and political life in the city. See Sánchez Korrol, *From colonia to community*. A useful early account is provided by Lawrence R. Chenault, *The Puerto Rican migrant in New York City* (New York: Columbia University Press, 1938). See also Arturo Morales Carrion, ed., *Puerto Rico: a political and cultural history* (New York: W. W. Norton, 1983); Manuel Maldonado-Denis, *Puerto Rico: a socio-historic interpretation* (New York: Random House, 1972); Fernando Picó, *Historia general de Puerto Rico* (Rio Pedras, Puerto Rico: Ediciones Huracán, 1986).

17

In many respects Operation Bootstrap represented one of the first attempts at "forcible modernization" in the developing world, and the initiative was strongly backed by a coterie of US academic, political, and economic interests. See Raymond Carr, *Puerto Rico: a colonial experiment* (New York: New York University Press, 1984); Thomas Mathews, *Puerto Rican politics and the New Deal* (Gainesville: University of Florida Press, 1960); and Berman Santana, "Geographers, colonialism, and development strategies."

18

Postwar migration patterns are reviewed in José Hernández Alvarez, "The movement and settlement of Puerto Rican migrants within the United States, 1950–1960," *International Migration*

Review 2 (1968): 40–51; Francesco Cordasco, Eugene Bucchioni, and Diego Castellanos, *Puerto Ricans on the United States mainland: a bibliography of reports, texts, critical studies and related materials* (Totowa, N.J.: Rowman and Littlefield, 1972); Francesco Cordasco and R. Galatioto, "Ethnic displacement in the interstitial community: the East Harlem experience," in Francesco Cordasco and Eugene Bucchioni, eds., *The Puerto Rican experience: a sociological sourcebook* (Totowa, N.J.: Rowman and Littlefield, 1973), pp. 171–185; Robert Novak, "Distribution of Puerto Ricans on Manhattan Island," *Geographical Review* (1956), pp. 18–36; Terry J. Rosenberg, *Residence, employment, and mobility of Puerto Ricans in New York City* (University of Chicago Department of Geography Research Paper no. 151, 1974).

19

Elena Padilla, *Up from Puerto Rico* (New York: Columbia University Press, 1958), pp. 301–308.

20

See Nathan Glazer and Daniel Patrick Moynihan, *Beyond the melting pot: the Negroes, Puerto Ricans, Jews, Italians, and Irish of New York City* (Cambridge, Mass.: MIT Press, 1963); Christopher Rand, *The Puerto Ricans* (New York: Oxford University Press, 1958); Clarence Senior, *The Puerto Ricans: strangers—then neighbors* (Chicago: Quadrangle Books, 1961).

21

Zaragosa Vargas, "Rank and file: historical perspectives on Latino/a workers in the US," *Humboldt Journal of Social Relations* (1996), pp. 11–23.

22

Sherries Baver, "Puerto Rican politics in New York City: the post-World War II period," in Jennings and Rivera, eds., *Puerto Rican politics in urban America,* pp. 43–59; Sánchez Korrol, *From colonia to community;* Carlos Rodríguez-Fraticelli and Amílcar Tirado, "Notes towards a history of Puerto Rican community organizations in New York City," *Centro: Journal of the Centro de Estudios Puertorriqueños* 2, no. 6 (1990): 17–35.

23

Robert Brenner, "The economics of global turbulence: a special report on the world economy, 1950–98," *New Left Review* 229 (1998): 1–265.

24

Wagenheim, *A survey of the Puerto Ricans on the US mainland;* History Task Force, Center for Puerto Rican Studies, *Labor migration under capitalism.*

25

"Disorders erupt in East Harlem; mobs dispersed," *New York Times,* 24 July 1967; "US troops sent into Detroit; 19 dead; Johnson decries riots; new outbreak in East Harlem," *New York Times,* 25 July 1967.

26

"Puerto Rican story: a sensitive people erupt," *New York Times,* 26 July 1967. Typical views of Puerto Rican apolitical "docility" before the late 1960s include Oscar Lewis, *La vida: a Puerto Rican family in the culture of poverty—San Juan and New York* (New York: Random House, 1965); Oscar Handlin, *The newcomers: Negroes and Puerto Ricans in a changing metropolis* (Cambridge, Mass.: Harvard University Press, 1959); Senior, *The Puerto Ricans;* Julian Steward, ed., *People of Puerto Rico: a study in social anthropology* (Champaign: University of Illinois Press, 1957). Contemporary perspectives that challenged this stereotype include Jesús Colón, *A Puerto Rican in New York and other sketches* (New York: Mainstream Publishers, 1961). "Not one of the statistical studies and dramatic presentations," wrote Colón, "conveys the slightest idea of the significant historical heritage of the Puerto Rican people" (p. 10).

27

Young Lords activist Eduardo Figueroa, interviewed in "The Young Lords: ten years after" (*Horizons,* National Public Radio, 1981), written, produced, and narrated by Elizabeth Tédas Muna. Tapes held in the Centro de Estudios Puertorriqueños at the City University of New York.

28

At its peak of organizational strength in the early 1970s, the Young Lords had branches in Chicago, Boston, Cleveland, Hayward (California), Newark and Hoboken (New Jersey), El Barrio, Bronx, and the Lower East Side (New York), Philadelphia, Bridgeport (Connecticut), Milwaukee, Hawaii, as well as Ponce, San Juan, and Aguadilla in Puerto Rico. By 1970 the Young Lords were one of the most militant and widespread of the Puerto Rican nationalist movements in the United States, their main rivals being the MLN and the FALN. Useful sources on the origins of the Young Lords include Frank Browning, "From rumble to revolution: the Young Lords," *Ramparts* (October 1970), reprinted in Cordasco and Bucchioni, eds., *The Puerto Rican experience,* pp. 231–245; Catarino Garza, ed., *Puerto Ricans in the US: the struggle for freedom* (New York: Pathfinder Press, 1977); *The Movement* (May 1969), p. 4; Augustín Laó, "Resources of hope: imagining the Young Lords and the politics of memory," *Centro: Journal of the Centro de Estudios Puertorriqueños* 7, no. 1 (1995): 34–49; Felipe Luciano, "The Young Lords Party, 1969–1975," *Caribe* 7 (1983) (New York: Visual Arts Center Relating to the Caribbean); Hilda Vasquez Ignatin, "Young Lords serve-and-protect," *Palante* 1, no. 2 (1969): 6; Jay Anthony Sanchez, "The Young Lords: past, present and promise" (BA thesis, Harvard University, 1989); Felix M. Padilla, *Puerto Rican Chicago* (Notre Dame, Indiana: University of Notre Dame Press, 1987), pp. 52–53, 121–122; and Carmen Teresa Whalen, "Bridging homeland and barrio politics: the Young Lords in Philadelphia," in Andrés Torres and José E. Velázquez, eds., *The Puerto Rican movement: voices from the diaspora* (Philadelphia: Temple University Press, 1998), pp. 107–123.

29

"Black Panthers join coalition with Puerto Rican and Appalachain groups," *New York Times,* 9 November 1969. Other radical groups operating at this time include the SDS Weathermen, Revolutionary Force 9, and the Appalachian Young Patriots. Other radical roots for the Young Lords within New York include the Real Great Society and Barrio Nuevo.

30

See Pablo "Yorúba" Guzmán, "La vida pura: a lord of the barrio," *Village Voice* (21 March 1995), pp. 24–31; Pablo "Yorúba" Guzmán, "Puerto Rican barrio politics in the United States," in Clara E. Rodriguez, Virginia Sánchez Korrol, and Oscar Alers, eds., *The Puerto Rican struggle: essays on survival in the US* (Maplewood, N.J.: Waterfront Press, 1984), pp. 121–128; Pablo "Yorúba" Guzmán, "Ain't no party like the one we got: the Young Lords and *Palante,*" in Ken Wachsberger, ed., *Voices from the underground: insider histories of the Vietnam era underground press* (Tempe, Ariz.: MICA's Press, 1993); "To the point . . . of production," *Palante* 1, no. 3 (1969): 8–9.

31

"Marchan en pos de justicia en welfare," *Palante* 1, no. 2 (1969): 5; Luis Garden-Acosta, former activist in the Young Lords and now director of the El Puente Academy for Peace and Justice in Brooklyn, New York, interview with the author (11 September 1997). The Sociedad de Albizu Campos had emerged in the late 1960s out of student activism within the state university system, principally at the Stony Brook and Old Westbury colleges, where many of the early leaders of the New York Young Lords began to develop their political ideas. The name was taken from the president of the Puerto Rican Nationalist Party who led the independence movement from the 1920s onward.

32

Pablo "Yorúba" Guzmán, cited in Abramson/Young Lords Party, *Palante: Young Lords Party,* p. 75.

33

Luis Garden-Acosta, interview with the author (11 September 1997).

34

On the cultural and political exchanges between Latinos and African-Americans see Juan Flores, *Divided borders: essays on Puerto Rican identity* (Houston, 1993), and William W. Sales, Jr., and Roderick Bush, "Black and Latino coalitions: prospects for new social movements in New York City," in James Jennings, ed., *Race and politics: new challenges and responses for black activism* (London: Verso, 1997), pp. 135–148. For some campaigns the Lords worked directly with the Black Power movement. In the autumn of 1970, for example, the Young Lords teamed up with Black Panthers and other groups to organize mass rent strikes against crumbling unheated tenements, demanding new boilers and pipes before winter. "Lords and Panthers organize rent strike," *New York Times,* 8 November 1970.

35

Cited in *Palante* 1, no. 4 (1969): 19.

36

Pablo "Yorúba" Guzmán, cited in Abramson/Young Lords Party, *Palante: Young Lords Party,* p. 75.

37

"Garbage burned in Harlem melee: trash thrown into streets to protest pick up service," *New York Times,* 18 August 1969.

38

Felipe Luciano, cited in *Palante* 1, no. 4 (1969): 17. See also "El Barrio and the YLO say no more garbage in our community," *Palante* 1, no. 4 (1969): 19; "La lucha contra la basura," *Palante* 1, no. 4 (1969): 20.

39

Dan Wakefield, *Island in the city: the world of Spanish Harlem* (Cambridge: Riverside Press, 1959), p. 232.

40

Estimates suggest that in the late 1960s only 30 percent of Puerto Ricans eligible to vote were actually registered, suggesting intense levels of alienation from electoral politics (a situation worsened by the English literacy tests used before the enactment of the 1965 Civil Rights Act). See Judith F. Herbstein, "Rituals and politics of the Puerto Rican 'community,' in New York City" (PhD dissertation, City University of New York, 1978).

41

"Puerto Rican group seizes church in East Harlem in demand for space," *New York Times,* 29 December 1969.

42

Luis Garden-Acosta, interview with the author (11 September 1997).

43

"Young Lords defy take over order," *New York Times,* 3 January 1970; "Militants vow to continue protest at Harlem church," *New York Times,* 4 January 1970; "Job loss laid to backing Young Lords," *New York Times,* 7 January 1970.

44

"Young Lords give food and care at seized church," *New York Times,* 30 December 1969. Other literature widely read by the Young Lords at their education classes include the works of S. Firestone, Franz Fanon, and Pablo Ferere. The provision of free breakfast for children was pioneered by the Black Panthers. See Miriam Eve White, "The Black Panthers' free breakfast for children program" (MA thesis, University of Wisconsin-Madison, 1988).

45

Luis Garden-Acosta, interview with the author (11 September 1997).

46

"105 members of Young Lords submit to arrest, ending 11-day occupation of church in East Harlem," *New York Times,* 8 January 1970.

47

On the prevalence of tuberculosis see Chenault, *The Puerto Rican migrant in New York City;* Abram J. Jaffe, *Puerto Rican population of New York City* (1954; New York: Arno Press, 1975). The political dimensions to interwar tuberculosis campaigns are explored in Nelida Perez, "A community at risk: Puerto Ricans and health, East Harlem, 1929–1940," *Centro: Journal of the Centro de Estudios Puertorriqueños* 1 (1988): 16–27, from which Eleanor Roosevelt is cited on p. 19.

48

Gloria Gonzales, cited in Abramson/Young Lords Party, *Palante: Young Lords Party,* p. 70. See also "Camion de placas-liberado," *Palante* 2, no. 6 (1970); "Tuberculosis mata/socialist medicine," *Palante* 2, no. 4 (1970): 8.

49

See, for example, Vivienne Bennett, *The politics of water: urban protest, gender, and power in Monterrey, Mexico* (Pittsburgh: University of Pittsburgh Press, 1995); Castells, *The city and the grassroots,* p. 68; Maureen Flanagan, "The city profitable, the city livable: environmental policy, gender, and power in Chicago in the 1910s," *Journal of Urban History* 22 (1996): 163–190; Angela Gugliotta, "Class, gender, and coal smoke: gender ideology and environmental injustice in

Pittsburgh, 1868–1914," *Environmental History* 5 (2000): 165–193; Harold L. Platt, "Jane Addams and the ward boss revisited: class, politics, and public health in Chicago, 1890–1930," *Environmental History* 5 (2000): 194–222.

50

By 1970 two women, Denise Oliver and Gloria Gonzales, had taken leadership positions in the party's Central Committee. Gonzales was to play a key role in the Lords' public health campaigns, including the Lincoln Hospital protests and the founding of HRUM (the Health Revolutionary Unity Movement) and its newspaper *For the People's Health.* In 1971 the Young Lords' Women's Caucus began its own campaigns and set up a Women's Union in East Harlem with a membership of between 50 and 70 people with its own newsletter, *La Luchadora.* See "Young women find a place in high command of Young Lords," *New York Times,* 11 November 1970; "La posición del partido de los Young Lords en cuanto a las mujeres," *Palante* 2, no. 13 (1970): 11–14; "Position on women's liberation," *Palante* 3, no. 8 (1971): 16–17; Sanchez, "The Young Lords." Other useful sources on Puerto Rican feminism include Y. Azize-Vargas, "The roots of Puerto Rican feminism: the struggle for universal suffrage," *Radical America* 23, no. 1 (1990): 71–79; I. Zavala Martínez, "En la lucha: the economic and socio-economic struggles of Puerto Rican women," in L. Fulani, ed., *The psyschopathology of everyday racism and sexism* (New York: Harrington Park, 1988); and Altagracia Ortiz, "The lives of pioneras: bibliographic and research sources on Puerto Rican women in the United States," *Centro: Journal of the Centro de Estudios Puertorriqueños* 2, no. 7 (1990): 40–47.

51

"Young Lords Party position paper on women," *Palante* 2, no. 12 (1970): 11–14.

52

A study of neighboring Newark, New Jersey, in 1974 found that of 200 Puerto Rican mothers only 12 had had any prenatal care. See Wagenheim, *A survey of the Puerto Ricans on the US mainland,* p. 52.

53

"Young Lords seize Lincoln Hospital building," *New York Times,* 15 July 1970.

54

Iris Morales, former activist in the Young Lords, director of the New York Network for School Renewal, interview with the author (11 September 1997).

55

Cited in Tugwell, *The stricken land,* p. 35.

56

Glazer and Moynihan, *Beyond the melting pot,* p. 117.

57

See Berman Santana, "Geographers, colonialism, and development strategies"; Peta Murray Henderson, "Population policy, social structure and the health system in Puerto Rico: the case of female sterilization" (PhD dissertation, University of Connecticut, 1976); M. Ladd-Taylor, "Saving babies and sterilizing mothers: eugenics and welfare politics in the inter-war United States," *Social Politics* 4, no. 1 (1997): 136–153; Iris Ofelia Lopez, "Sterilization among Puerto Rican women: a case study in New York City" (PhD dissertation, Columbia University, 1985); B. Maas, *Population target: the political economy of population control in Latin America* (Toronto: Latin American Working Group, 1976); H. Presser, *Sterilization and fertility in Puerto Rico* (Berkeley: Institute of International Studies, University of California, 1973); José Vasquez Calzada, "La esterilización femenina en Puerto Rico," *Revista de Ciencias Sociales* 17, no. 3 (1973): 283–300; and "Sterilized Puerto Ricans," *Palante* 2, no. 10 (1970): 5, 20.

58

Altagracia Ortiz, "'En la aguja y el pedal eché la hiel': Puerto Rican women in the garment industry of New York City, 1920–1980," in Altagracia Ortiz, ed., *Puerto Rican women and work: bridges in transnational labor* (Philadelphia: Temple University Press, 1966), pp. 55–81.

59

See José E. Cruz, *Identity and power: Puerto Rican politics and the challenge of ethnicity* (Philadelphia: Temple University Press, 1988); Roberto P. Rodríguez-Morazzani, "Puerto Rican political generations in New York: *pioneros,* young Turks and radicals," *Centro: Journal of the Centro de Estudios Puertorriqueños* 4, no. 1 (1992): 96–116; Juan Angel Silén, *We, the Puerto Rican people: a story of oppression and resistance,* trans. Cedric Belfrage (1970; London: Monthly Review Press, 1971).

60

See Elaine Brown, *A taste of power: a black woman's story* (New York: Anchor Press, 1994); Aida Hurtado, *The color of privilege: three blasphemies on race and feminism* (Ann Arbor: University of Michigan Press, 1996); Marisela Rodríguez Chávez, "Living and breathing the movement: women in *el centro de acción social autónomo,* 1975–1978" (MA thesis, Arizona State University, 1997); Ann Standley, "The role of black women in the civil rights movement," in Vicki L. Crawford, Jacqueline Ann Rose, and Barbara Woods, eds., *Women in the civil rights movement: trailblazers and torchbearers, 1941–1965* (Bloomington: Indiana University Press, 1990), pp. 183–202.

61

Gil Scott-Heron, *Winter in America* (New York: Brouhaha Music, 1975).

62

Denise Oliver, former Young Lords activist, interviewed in "The Young Lords: ten years after" (*Horizons,* National Public Radio, 1981).

63

Dardo Cúneo, cited in Herbstein, "Rituals and politics of the Puerto Rican 'community,'" p. 30.

64

Efraín Barradas, cited in J. Jorge Klor de Alva, "Aztlán, Borinquen, and Hispanic Nationalism in the United States," in Rudolfo A. Anaya and Francisco A. Lomeli, eds., *Aztlán: essays on the Chicano homeland* (Albuquerque: University of New Mexico Press, 1989), pp. 135–181.

65

Luis Garden-Acosta, interview with the author (11 September 1997). The Young Lords' control over street violence became precarious. In June 1970, for example, hundreds of youths were involved in looting and rioting between 105th and 116th streets in East Harlem. In October 1974 five bombs went off in Manhattan and nearly 20,000 people gathered at Madison Square Garden for a Puerto Rican independence rally. Between 1974 and 1984 there were some 40 unsolved bombings in New York City that are thought to be connected with Puerto Rican separatist groups such as the Fuerzas Armadas de Liberación Nacional (FALN).

66

Herbstein, "Rituals and politics of the Puerto Rican 'community.'"

67

The early 1970s also saw increasing political polarization and agitation in Puerto Rico. A pro-independence rally in San Juan attracted a crowd of 80,000 who protested against the drafting of Puerto Ricans to serve in the Vietnam War, US navy exercises in nearby Culebra, and the expansion of the onshore activities of environmentally damaging US oil companies. See Garza, ed., *Puerto Ricans in the US.*

68

On the complexity of race as a basis for political mobilization in the Puerto Rican community see Antonia Darder and Rodolfo D. Torres, "Latinos and society: culture, politics, and class," in Antonia Darder and Rodolfo D. Torres, eds., *The Latino studies reader: culture, economy & society* (Oxford and Malden, Mass.: Blackwell, 1998), pp. 3–26; Juan Flores, "Pan-Latino/trans-Latino: Puerto Ricans in the 'New Nueva York,'" *Centro: Journal of the Centro de Estudios Puer-*

torriqueños 8, nos. 1–2 (1996): 171–186; Roberto P. Rodríguez-Morazzani, "Beyond the rainbow: mapping the discourse on Puerto Rican and 'race,'" *Centro: Journal of the Centro de Estudios Puertorriqueños* 8, nos. 1–2 (1996): 151–169.

69

Some indication of the seriousness with which American security forces treated the Young Lords is suggested by an extensive report produced in 1976 for the US Senate. We can trace a continuum of FBI involvement between the 1930s in Puerto Rico and urban American protest in the 1970s. The significance of FBI infiltration is also raised in Padilla, *Puerto Rican Chicago,* p. 173.

70

Luis Garden-Acosta, interview with the author (11 September 1997).

71

Wakefield, *Island in the city;* Herbert Hill, "Guardians of the sweatshops: the trade unions, racism and the garment industry," in Adalberto López and James Petras, eds., *Puerto Rico and Puerto Ricans: studies in history and society* (Cambridge, Mass.: Schenkman, 1974), pp. 384–417.

72

Mike Davis, *Prisoners of the American dream: politics and economy in the history of the US working class* (London: Verso, 1986); Ira Katznelson, *City trenches: urban politics and the patterning of class in the United States* (Chicago: University of Chicago Press, 1981). The principal challenge, as Ira Katznelson observes, is that "community-based strategies for social change in the United States cannot succeed unless they pay attention to the country's special pattern of class formation; to the split in the practical consciousness of American workers between the language and practice of a politics of work and those of a politics of community" (p. 194). In a sense, then, the Young Lords' organizational strategy pushed this dilemma to its limit through their radical focus on the ghetto experience as a means to foster political consciousness.

73

On the return of gang violence see, for example, "Upsurge of gang violence in Brooklyn," *New York Times,* 29 June 1973, p. 39.

74

Andres Torres and Frank Bonilla, "Decline within decline: the New York perspective," in Morales and Bonilla, *Borderless borders.*

75

Despite the demise of the Young Lords in the early 1970s, many of the activists have gone on to play significant leadership roles in the Puerto Rican community in the 1980s and 1990s.

Examples include David Perez (professor of Puerto Rican history at Hunter College), Iris Morales (trade union lawyer and documentary filmmaker, director of education at the Puerto Rican Legal Defense and Education Fund), Minerva Solla (trade union organizer), Luis Garden-Acosta (director of the radical community college and environmental education center El Puente), Juan González (journalist), Pablo Guzmán (journalist), Richie Pérez (national coordinator of the National Congress for Puerto Rican Rights), Julio Pabon (leader of the Coalition in Defense of Puerto Rican and Hispanic Rights). For some indication of the continuing significance of the Young Lords legacy see "The Lords of East Harlem," *New York Newsday,* 24 January 1990, and "Young Lords: vital in 60's, a model now," *New York Times,* 16 October 1996.

76

José F. Morales, "The toxic avengers," *Centro: Journal of the Centro de Estudios Puertorriqueños* 2, no. 5 (1990): 109–114.

77

Déborah Berman Santana, *Kicking off the bootstraps: environment, development, and community power in Puerto Rico* (Tucson: University of Arizona Press, 1996); Centro de Información, Investigación y Educación Social (CIIES), *La lucha ambiental en Puerto Rico: un poco de historia* (Hato Rey: CIIES, 1992); Neftalí García Martínez, *¿Quién cantará por las aves? Ensayos sobre el ambiente puertorriqueño* (San Juan: Servicios Científicos y Técnicos, 1996); Marya Muñoz-Vázquez, "Gender and politics: grassroots leadership among Puerto Rican women in a health struggle," in Ortiz, ed., *Puerto Rican women and work,* pp. 161–183; US/Puerto Rico Solidarity Network, "Environmental fronts of struggle: an interview with Dr. Neftalí García Martínez," 1991 (copy held in the Centro de Estudios Puertorriqueños at the City University of New York). Other key organizations in the Puerto Rican environmental movement include the Centro de Acción Ambiental (Environmental Action Center); Frente Loiceños Unidos (United Loíza Front); Comité Pro Rescate del Buen Ambiente de Guayanilla (Committee to Save the Environment in Guayanilla); Sur Contra la Contaminación (South Against Pollution); and Comité Pro Rescate y Desarrollo de Vieques (Committee to Rescue and Develop Vieques).

78

For more details on Chicano environmental activism see, for example, Benjamin Marquez, "Mobilizing for environmental and economic justice: the Mexican-American environmental justice movement," *Capitalism, Nature, Socialism* 9 (1998): 43–60; Laura Pulido, *Environmentalism and economic justice: two Chicano struggles in the Southwest* (Tucson: University of Arizona Press, 1996).

79

Katznelson, *City trenches,* p. 212.

80

Castells, *The city and the grassroots,* pp. 318–327.

81

Ibid., p. 320.

82

See Guy-Ernest Debord, "Introduction à une critique de la géographie urbaine," in Gérard Berreby, ed., *Documents relatifs à la fondation de l'Internationale situationniste* (Paris: Allia, 1995); Greil Marcus, *Lipstick traces: a secret history of the twentieth century* (Cambridge, Mass.: Harvard University Press, 1989).

5

RUSTBELT ECOLOGY

1

Giuliana Bruno, "Ramble city: postmodernism and *Blade Runner,*" *October* 41 (1987): 61–84.

2

Don DeLillo, *Underworld* (New York: Scribner, 1997), pp. 183–184.

3

Andrew White, "Small beginnings: the birth of environmental activism in Fort Greene's housing projects," *City Limits* (April 1992), p. 14.

4

On the rustbelt-to-sunbelt shift see Barry Bluestone and Bennett Harrison, *The deindustrialization of America* (New York: Basic Books, 1982); Bennett Harrison and Barry Bluestone, *The great U-turn: corporate restructuring and the polarizing of America* (New York: Basic Books, 1988); Larry Sawers and William K. Tabb, eds., *Sunbelt/snowbelt* (Oxford: Oxford University Press, 1984). On the geography of "new industrial spaces" see A. J. Scott, *New industrial spaces: flexible production, organisation and regional development in North America and Western Europe* (London: Pion, 1988). For the postwar economic restructuring of the New York metropolitan region see, for example, Matthew P. Drennan, "The decline and rise of the New York economy," in John H. Mollenkopf and Manuel Castells, eds., *Dual city: restructuring New York* (New York: Russell Sage

Foundation, 1991), pp. 25–41; Beverly Duncan and Stanley Liberson, *Metropolis and region in transition* (Beverley Hills, Calif.: Sage, 1970); Robert Fitch, *The assassination of New York* (London: Verso, 1993); Michael K. Heiman, *The quiet evolution: power, planning, and profits in New York State* (New York: Praeger, 1988).

5

Rosalyn Deutsche, "The threshole of democracy," in Bronx Museum of the Arts, *Urban mythologies: the Bronx represented since the 1960s* (New York: Bronx Museum of the Arts, 1999), p. 97.

6

See, for example, Steve Mitra, "Clean sweep: neighborhoods want to be rid of private industry's garbage transfer stations," *City Limits* (April 1994), pp. 14–15; Margaret Mittelbach, "The dumping fields," *City Limits* (January 1992), pp. 14–18.

7

The significance of waste paper in the New York City urban ecosystem was noted by David Harvey, "Dialectics, environmental and social change," lecture given at the Institute for Historical Research, Senate House, University of London (11 February 1992).

8

City of New York Independent Budget Office, "Inside the budget: a newsfax of the Independent Budget Office. No. 65. City proposes garbage export plan" (17 July 2000).

9

For an examination of the political and historical dimensions to different waste disposal strategies, see Louis Blumberg and Robert Gottlieb, *War on waste: can America win its battle with garbage?* (Washington, D.C.: Island Press, 1989); Harold Crooks, *Giants of garbage: the rise of the global waste industry and the politics of pollution control* (Toronto: James Lorimer, 1993); Elizabeth Fee and Steven H. Corey, *Garbage! The history and politics of trash in New York City* (New York: New York Public Library, 1994); Hildegard Frilling and Olaf Mischer, *Pütt un Pann'n: Geschichte der Hamburger Hausmüllbeseitigung* (Hamburg: Ergebnisse, 1994); Matthew Gandy, *Recycling and the politics of urban waste* (New York: St. Martin's Press, 1994); Institut für ökologisches Recycling, *Ökologische Abfallwirtschaft: Umweltvorsage durch Abfallvermeidung* (Berlin: IföR, 1989); Martin Melosi, *Garbage in the cities: refuse, reform, and the environment, 1880–1980* (College Station: Texas A & M University Press, 1981); Susan Strasser, *Waste and want: a social history of trash* (New York: Metropolitan Books, 2000).

10

For a critique of recycling and Western environmentalism see Stephen Horton, "Rethinking recycling: the politics of the waste crisis," *Capitalism, Nature, Socialism* 6 (1995): 1–19; Timothy

J. Luke, "Green consumerism: ecology and the ruse of recycling," in Jane Bennett and William Chaloupka, eds., *In the nature of things: language, politics, and the environment* (Minneapolis: University of Minnesota Press, 1993), pp. 154–172. See also Barry Castleman and Vicente Navarro, "International mobility of hazardous products, industries, and wastes," *Annual Review of Public Health* 8 (1987): 1–19.

11

Fee and Corey, *Garbage!* See also Jacob A. Riis, *How the other half lives* (London: Sampson, Low, Marston, Searle, & Rivington, 1891), pp. 50–52.

12

George E. Waring, "The disposal of a city's waste," *North American Review* (July 1895), p. 49.

13

Horton, "Rethinking recycling"; Simone Martzloff, "Perspektiven der Abfallvermeidung," in *Perspektive Abfallvermeidung* (Berlin: Institut für ökologisches Recycling, 1991), pp. 9–16. An influential critique of American consumerism is provided by Vance Packard, *The waste makers* (1960; Harmondsworth: Pelican Books, 1963).

14

Paul Bradley, "Garbage wars," *City Limits* (January 1998), pp. 18–23. A study commissioned by the Department of Sanitation estimated that the city produces over 24,000 tons per day of municipal solid waste, including household, commercial, and institutional wastes, but excluding medical wastes, construction wastes, and sewage sludge. Tellus Institute, *A statistical profile of New York City for solid waste management planning* (1991).

15

One closed section of the Fresh Kills landfill stands at 210 feet, compared with 204 feet for the Mycerinus (Menkaure) Pyramid, while it is estimated that by the year 2010 one active section of the site will reach 450 feet, exceeding the height of the King Kephren Pyramid. Interview with Nick Dmytryszyn, Environmental Engineer to the Staten Island Borough President (18 October 1993). Information on the Egyptian pyramids was obtained from the Egyptian Tourist Office, London.

16

Charles Arnold, Assistant to the Director of Recycling Implementation, New York City Department of Sanitation, interview with the author (26 August 1993). Nancy Wolf and Steve Richardson, Environmental Action Coalition, New York City, interviews with the author (26 August 1993). The decision to close Fresh Kills in 2001 is for primarily political reasons, since it still has a technical lifespan of twenty years. An interim contract has been signed to incinerate waste in Newark, New Jersey, after the closure of Fresh Kills, which is ironic in view of the defeat of incineration plans in Brooklyn partly on the pretext that incinerators in New Jersey already present a health threat to New Yorkers. Nancy Anderson, City of New York, Office of Policy Management, personal communication with the author (3 November 1999).

17

"Environmental concerns unite a neighborhood," *New York Times,* 15 January 1993.

18

The reference to Division Avenue is taken from Randy Shaw, *The activist's handbook: a primer for the 1990s and beyond* (Berkeley: University of California Press, 1996), p. 84.

19

"Hasidic and Hispanic residents in Williamsburg try to forge a new unity," *New York Times,* 18 September 1994; John Fleming, La Puente, New York City, interview with the author (9 March 1995); Suzanne Mattei, New York City Office of the Comptroller, interview with the author (3 March 1995).

20

Shaw, *The activist's handbook,* p. 84. In 1993 El Puente gained status as one of the city's new theme-oriented public schools, operating in partnership with the Board of Education. The El Puente Academy for Peace and Justice now includes a dance ensemble, a muralists' group, a video documentary project, a health clinic, an AIDS prevention program, and a youth environmentalist group "A bridge from hope to social action," *New York Times,* 23 May 1995. Luis Garden-Acosta, former Young Lords activist and Director of the El Puente Academy for Peace and Justice, New York, interview with the author (11 September 1997).

21

K. Greider, "Against all odds: after years of interethnic turmoil in North Brooklyn, Community Alliance for the Environment is building bridges between Latinos and Hasidim," *City Limits* (August/September 1993), p. 34; "Fear builds bridge across gulf of cultures," *New York Times,* 31 August 1992. On early protests against the plant see Ellen Kissman, "The role of risk in the siting of public facilities: dioxins and the Brooklyn Navy Yard resource recovery plant" (MSc thesis, Columbia University, 1987). Opposition to the Radiac Corporation was an early source of dialogue between the Latino and Hasidic communities. In the late 1980s, for example, a series of Latino youth-led groups such as the Toxic Avengers, Earth Spirit, and Brooklyn Urban Mines (BUMs) played a key role in politicizing waste issues. The Toxic Avengers, who disbanded in the early 1990s, were named after the cult film *The Toxic Avenger* (1984) directed by Michael Herz, a satirical portrayal of the "toxic waste capital" of America called Tromaville in

New Jersey. Alicia Fierer, Environmental Action Coalition, New York City, interview with the author (1 March 1995); John Fleming, interview with the author (22 March 1995); M. Holloway, "The toxic avengers take Brooklyn," *City Limits* (December 1989), pp. 8–9.

22

Greider, "Against all odds," p. 34.

23

Greider, "Against all odds," p. 35; Community Environmental Health Center (CEHC), *Hazardous neighbors? Living next door to industry in Greenpoint-Williamsburg* (New York: Hunter College, City University of New York, 1989); Steve Mitra, "Toxic revenge," *City Limits* (June/July 1994), pp. 16–18.

24

White, "Small beginnings."

25

Willine Carr, Lisa Zeitel, and Kevin Weiss, "Variations in asthma hospitalizations and deaths in New York City," *American Journal of Public Health* 82 (1992): 59–65; Douglas W. Dockery, "An association between air pollution and mortality in six US cities," *New England Journal of Medicine* 329 (1993): 1753–1759; R. Evans, "Asthma among minority children: a growing problem," *Chest* 101 (1992): 368–371; D. J. Gottlieb, A. S. Beiser, and G. T. O'Connor, "Poverty, race and medication use are correlates of asthma hospitalization rates: a small area analysis in Boston," *Chest* 108 (1995): 28–35; R. D. Morris, E. N. Naumova, and R. L. Munisinghe, "Ambient air pollution and hospitalization for congestive heart failure among elderly people in 7 large US cities," *American Journal of Public Health* 85 (1995): 1361–1365. See also Agency for Toxic Substances Disease Registry, *The nature and extent of lead poisoning in children in the United States: a report to Congress* (Atlanta: US Department of Health and Human Resources, 1988); K. Sexton, H. Gong, J. C. Bailar, B. Bryant, J. G. Ford, D. R. Gold, W. E. Lambert, and M. J. Utell, "Air pollution health risk: do class and race matter?," *Toxicology and Industrial Health* 9 (1993): 843–878; John Walsh, "Unhealthy atmosphere in Harlem," *Science* (17 January 1989), pp. 269–278; "Soot city: Manhattan a hot spot for deadly pollution," *New York Newsday,* 9 December 1993; "Asthma common and on the rise in the crowded South Bronx," *New York Times,* 5 September 1995.

26

See, for example, Karen Brudney and Jay Dobkin, "Resurgent tuberculosis in New York City: human immunodeficiency virus, homelessness, and the decline of tuberculosis control programs," *American Review of Respiratory Disease* 144 (1991): 745–749; E. Drucker, P. Alcabes,

W. Bosworth, and B. Sckell, "Childhood tuberculosis in the Bronx, New York," *Lancet* 343 (1994): 1482–1485; Roderick Wallace, Deborah Wallace, H. Andrews, R. Fullilove, and M. T. Fullilove, "The spatiotemporal dynamics of AIDS and TB in the New York metropolitan region from a sociogeographic perspective: understanding the linkages of central city and suburbs," *Environment and Planning A* 27 (1995): 1085–1108; Roderick Wallace and Deborah Wallace, "The destruction of US minority urban communities and the resurgence of tuberculosis: ecosystem dynamics and the white plague in the dedeveloping world," *Environment and Planning A* 29 (1997): 269–291.

27

Cited in Bradley, "Garbage wars," p. 20.

28

Arnold Markoe, "Brooklyn Navy Yard," in Kenneth T. Jackson, ed., *The encyclopedia of New York City* (New Haven: Yale University Press, 1995), pp. 159–160; Marshall Berman, *All that is solid melts into air: the experience of modernity* (London: Verso, 1982), p. 325; "Williamsburg; a new battle over Navy Yard," *New York Times,* 10 October 1993.

29

The city's earliest example of the use of incineration for the production of electricity can be traced to 1905, with the construction of a plant that provided lighting for bridges and highways. By the late 1930s the La Guardia administration had built two incinerators equipped with turbine generators for the production of electricity, but the New York Edison Company refused to buy the electricity, making incineration uneconomical in comparison with waste disposal by landfill.

30

See Fee and Corey, *Garbage!;* City of New York, *Freedom to breathe: report of the mayor's task force on air pollution in the City of New York* (New York: City of New York, 1966); S. Tannenbaum, "A brief history of waste disposal in New York City since 1930," unpublished paper held in the New York City Municipal Archives, 1992; George A. Soper, *Modern methods of street cleaning* (London: Archibald Constable and Co., 1909), pp. 161–162; Ernest P. Goodrich, "The opportunities for refuse salvage in New York City," *American City* (February 1935), p. 58.

31

The growing vulnerability of the city's waste disposal system to the loss of the Fresh Kills landfill site can be illustrated by the impact of the three-month tugboat strike of 1979, which prevented the city from using its barge transport system to deliver waste to Staten Island. The consequences of this period of disruption were far reaching: additional costs incurred by the city exceeded $4 million; the greater travel distances and congestion at alternative landfill sites increased the time taken for delivering each load by up to three hours; in order to maintain adequate collection services under these circumstances, street cleaning had to be curtailed and personnel reassigned to waste disposal operations; mandatory overtime had to be instituted to handle the increased work load; there was a greater need for repair and maintenance work on the overworked equipment; there was a marked increase in gasoline consumption because of the reliance on road haulage; and finally, the streets became dirtier and a source of concern for public health. See City of New York Department of Sanitation, "The waste disposal problem in New York City: a proposal for action" (submitted in April 1984), pp. 1–6; "Norman Steisel and the art of the done deal," *Village Voice* (26 November 1991). The Fresh Kills landfill also contains unknown quantities of hazardous wastes deposited there between its creation in 1948 and the more stringent environmental regulations of the 1970s. See Eric A. Goldstein and Mark A. Izeman, *The New York environment book* (Washington, D.C.: Island Press, 1990); M. de Kadt and N. Lilienthal, *Solid waste management: the garbage challenge for New York City* (New York: INFORM, 1989); City of New York Office of the Comptroller, "Burn, baby, burn: how to dispose of garbage by polluting land, sea and air at enormous cost" (submitted January 1992).

32

The key piece of federal legislation is the Public Utilities Regulatory Policy Act introduced by the Carter administration in 1978: now that this legislation has been amended, a key aspect to the economic case for incineration has been removed. City of New York Department of Sanitation, "The waste disposal problem in New York City"; City of New York Department of Sanitation, "Comprehensive solid waste management plan for New York City and final generic environmental impact statement" (submitted in August 1992).

33

On the shift of engineering companies from nuclear energy to waste incineration see "Norman Steisel and the art of the done deal"; "World nuclear industry," *Financial Times Survey* (17 November 1993); Michael K. Heiman, "Up in smoke: the fate of solid waste incineration and the so-called garbage crisis," paper presented to the Annual Meeting of the Association of American Geographers, Honolulu, 24–27 March 1999.

34

Nevin Cohen, "Technical assistance for citizen participation: a case study of New York City environmental planning process," *American Review of Public Administration* 25 (1995): 119–135.

35

A good summary of the anti-incineration case in the 1980s is provided by Barry Commoner, "The folly of incineration," *City Limits* (June/July 1988), pp. 8–10; "$90 million garbage-fueled plant proposed for Brooklyn Navy Yard," *New York Times,* 20 December 1977; "Navy Yard is pushed as site for city's 1st trash recycling plant," *New York Post,* 5 March 1979; "City plan for garbage fueled plant draws anger of Brooklyn residents," *New York Times,* 31 May 1981; "Approval is given for incinerator at Brooklyn site," *New York Times,* 16 August 1985; "Board of Estimate approves plan for 5 incinerators," *New York Times,* 21 December 1984; "City's plan to build 8 incinerators is debated at first public hearing," *New York Times,* 3 May 1984; Charles Turner, "A waste-to-energy plan looms over Brooklyn," *City Limits* (November 1982), pp. 12–16. See the range of views expressed in the City of New York Department of Sanitation, "Final environmental impact statement for proposed resource recovery facility at the Brooklyn Navy Yard" (submitted June 1985); "Pollution threat is cited in study on incinerators; increase in lung cancer is seen from city's plan," *New York Times,* 30 April 1984; "Dioxin dispute continues as incinerator vote nears," *New York Times,* 5 December 1984; "As thousands protest, city backs trash plan," *New York Times,* 4 August 1985. The growing antipathy toward waste incineration was not restricted to the United States, with high profile campaigns also emerging in Western Europe where waste issues had also been extensively politicized in the 1980s. See Barbara Ettorre, "Are we headed for a recycling backlash," *Management Review* (June 1992), p. 14; E. Spill and E. Wingert, eds., *Brennpunkt Müll* (Hamburg: Sternbuch, 1990).

36

John Mollenkopf, *A phoenix in the ashes: the rise and fall of the Koch coalition in New York City politics* (Princeton, N.J.: Princeton University Press, 1992); William W. Sales, Jr., and Roderick Bush, "Black and Latino coalitions: prospects for new social movements in New York City," in James Jennings, ed., *Race and politics: new challenges and responses for black activism* (London: Verso, 1997), pp. 135–148.

37

Barbara Day, "New York: David Dinkins opens the door," in Mike Davis, Steven Hiatt, Marie Kennedy, Susan Ruddick, and Michael Sprinkler, eds., *Fire in the hearth: the radical politics of place in America* (London: Verso, 1990), pp. 159–175; Michelle de Pass, executive director, New York City Environmental Justice Alliance, interview with the author (4 April 1995); "Anti-incineration stand brings a blessing," *New York Times,* 18 October 1989.

38

"Community demands Dinkins cancel Brooklyn incinerator," *Amsterdam News,* 9 May 1992; Patricia Grayson, Bureau of Waste Prevention, Reuse and Recycling, New York City Department of Sanitation, interview with the author, 26 August 1993; "Dinkins burning for incinerator," *New York Newsday,* 26 August 1992; "Full steam ahead for garbage plan," *New York*

Newsday, 27 August 1992; "Race, poverty and the environment; pollution-weary minorities try civil rights tack," *New York Times,* 11 January 1993; "State approves the Brooklyn Navy Yard incinerator," *New York Times,* 12 September 1993; "Brooklyn incinerator opponents see victory in high court delay," *New York Times,* 4 May 1994; "New York State adds Navy Yard to list of hazardous waste sites," *New York Times,* 5 February 1995; "Plan to build incinerator faces delay," *New York Times,* 16 June 1995. In 1996 further objections to the project were raised because of the disturbance to slave burial grounds discovered on the site by archaeologists. See "Stir over thousands of African bones buried at Brooklyn Navy Yard," *Amsterdam News,* 23 March 1996. The calculations of the percentage of minorities among the population living within 4 kilometers of the proposed site is taken from City of New York Office of the Comptroller, "Smokescreen: how the Department of Sanitation's solid waste plan and environmental impact statement cover up the poisonous health effects of burning garbage" (1992).

39

See, for example, the pioneering work of Barry Commoner and his colleagues: B. Commoner, M. McNamara, K. Shapiro, and T. Webster, *Environmental and economic analysis of alternative municipal solid waste disposal technologies. I. An assessment of the risks due to emissions of chlorinated dioxins and dibenzofurans from proposed New York City incinerators* (Flushing, N.Y.: Center for the Biology of Natural Systems, 1984); B. Commoner, M. McNamara, K. Shapiro, and T. Webster, *Environmental and economic analysis of alternative municipal solid waste disposal technologies. III. A comparison of different estimates of the risks due to emissions of chlorinated dioxins and dibenzofurans from proposed New York City incinerators (including a critique of the Hart Report)* (Flushing, N.Y.: Center for the Biology of Natural Systems, 1984). On the Agent Orange connection see L. C. Casten, "Agent Orange's forgotten victims," *Nation* (4 November 1991). On the legacy of dioxins used in the Vietnam War see, for example, N. T. N. Phuong, B. S. Hung, A. Schechter, and D. Q. Vu, "Dioxin levels in adipose tissues of hospitalized women living in the south of Vietnam in 1984–1989 with a brief review of their clinical histories," *Women and Health* 16 (1990): 79–93.

40

Barry Commoner, interview with the author (20 June 1995). See also Barry Commoner, "The origin and health risks of PCDD and PCDF," *Waste Management & Research* 5 (1987): 327–346; "Scientists are unlocking secrets of dioxin's devastating power," *New York Times,* 15 May 1990.

41

City of New York Office of the Comptroller, "Burn, baby, burn."

42

Nancy Anderson, personal communication with the author (3 November 1999); City of New York Office of the Comptroller, "A tale of two incinerators: how New York City opposes incineration in New Jersey while supporting it at home" (May 1992).

43

New York State Department of Environmental Conservation, Division of Solid and Hazardous Waste, "Ash residue characterization report" (July 1987); Ron Davis and Barry Meier, "Will it go up in smoke? The New York experience," in Newsday, *Rush to burn: solving America's garbage crisis* (Washington, D.C.: Island Press, 1989), pp. 140–150.

44

City of New York Office of the Comptroller, "Burn, baby, burn"; Environmental Defense Fund, *To burn or not to burn: the economic advantages of recycling over garbage incineration for New York City* (New York: Environmental Defense Fund, 1985); New York Public Interest Research Group, *A fiscal analysis of the City of New York's solid waste management programs and the proposed Brooklyn Navy Yard incinerator* (New York: NYPIRG, 1992).

45

City of New York Office of the Comptroller, "Fire and ice: how garbage incineration contributes to global warming" (March 1993).

46

See, for example, the statement of Eric A. Goldstein on behalf of the Natural Resources Defense Council before the New York City Council Environmental Protection Committee concerning New York City's solid waste management plan, 25 August 1992.

47

"Only with Local Law #19 of 1989," write the historians Elizabeth Fee and Steven Corey, "did New York City again adopt an official recycling plan—fitfully returning, almost a hundred years later, to a project started in the 1890s." Fee and Corey, *Garbage!,* p. 54. The potential contribution of recycling to solving the city's waste management crisis is a highly contentious issue. Only aluminum commands a consistently strong price on the secondary materials market, and over half of the municipal waste stream consists of putrescible kitchen and yard wastes for which no viable market currently exists. Markets for glass and newsprint have weakened in the 1990s, complicating any long-term planning for the fiscal implications of postconsumer recycling in comparison with cheaper alternatives such as landfill. See, for example, Andrew White, "Disposable dreams," *City Limits* (October 1991), pp. 12–15; James Bradley, "Waste not, want not:

tons of material collected at the curbside is ending up in landfills," *City Limits* (May 1994), pp. 14–18; Paul Bradley, "Losing ground," *City Limits* (June/July 1995), pp. 14–18; "Study says recycling effort could fail in New York City," *New York Times,* 12 October 1995. On the failure of recycling in Los Angeles see Horton, "Rethinking recycling." By the autumn of 1993 all 59 of New York's Community Districts were integrated into a comprehensive mandatory recycling program reaching a recycling rate of around 14 percent, which compares favorably with the extent of recycling in other large metropolitan areas in developed economies. More optimistic assessments of the potential for recycling include Luke Cyphers, "The green alternative: a visionary plan for New York's garbage," *City Limits* (April 1992), pp. 10–11; Center for the Biology of Natural Systems, *Development of innovative procedures to achieve high rates of recycling in urban low-income neighborhoods,* report submitted to the Pew Charitable Trusts and the Aaron Diamond Foundation (Flushing, N.Y.: Center for the Biology of Natural Systems, Queens College, City University of New York, 1992); L. T. Hang and S. A. Romalewski, *The burning question: garbage incineration vs. total recycling in New York City* (New York: Public Interest Research Center, 1986); Environmental Defense Fund, *To burn or not to burn;* New York Public Interest Research Group, *A fiscal analysis.*

48

"City is given more time for goals on recycling," *New York Times,* 28 May 1997; "Giuliani attacks recycling goals as a suit is filed," *New York Times,* 3 July 1996; "Giuliani says recycling law is already met," *New York Times,* 4 July 1996. The city's 1998 *Annual report on social indicators* records that 1,124,136 tons of waste were recycled in 1998. It is unlikely, however, that this high figure refers only to municipal waste, which would imply a recycling rate of 37 percent for household waste. These revised figures are more likely to reflect a political change in the definition of recycling stemming from the city's failure to meet the mandatory 25 percent target. See also City of New York Independent Budget Office, "Inside the budget: City proposes garbage export plan."

49

See Melosi, *Garbage in the cities.*

50

A. Darnay and W. E. Franklin, *Salvage markets for materials in solid wastes* (Washington, D.C.: Environmental Protection Agency, 1972).

51

Horton, "Rethinking recycling," p. 3.

52

Thomas H. Garver, "Serving places," in Stanley Greenberg, *Invisible New York: the hidden infrastructure of the city* (Baltimore: Johns Hopkins University Press, 1998), p. 4.

53

White, "Small beginnings."

54

"Norman Steisel and the art of the done deal."

55

Shaw, *The activist's handbook,* pp. 87–88.

56

US Environmental Protection Agency Risk Ranking Work Group, Region II, *Overview report: comparative risk ranking of the health, ecological and welfare effects of twenty-seven environmental problem areas in Region II* (Washington, D.C.: Environmental Protection Agency, 1991.

57

City of New York Department of Sanitation, "Public comments on a comprehensive solid waste management plan for New York City and draft generic environmental impact statement" (submitted in June 1992).

58

"Dinkins delays building incinerator in Brooklyn," *New York Times,* 25 March 1992.

59

Michael K. Heiman notes that long-term debt financing or bond credit for incinerators is more complex than for other kinds of infrastructure projects because the quality and quantity of the waste stream is difficult to predict. Heiman, "Up in smoke"; see also "New York's garbage trucks and trash headed to New Jersey," *Big Apple Garbage Sentinel* (23 September 1999). BAGS can be viewed at http://garbagesentinel.org.

60

Michael K. Heiman, academic and community activist, personal communication with the author (16 December 1999).

61

"Garbage is one thing but garbage from New York? Forget it," *New York Times,* 12 February 1989.

62

I must thank one of my manuscript reviewers at MIT for this observation.

63

Blaine Harden, "Trade trash for culture? Not Virginia," *New York Times,* 18 January 1999.

64

"New York mulls dumping trash outside state: city contract could span two decades and inject $1 billion into market," *Wall Street Journal,* 4 March 1994. Nancy Anderson, personal communication with the author (3 November 1999).

64

Steven Corey, paper presented to the conference of the American Association of Environmental Historians, Tacoma, 16–19 March 2000.

66

"In South Bronx, a bitter split over a proposed transfer station," *New York Times,* 29 September 1999; "As deadline looms for dump, alternate plans prove elusive," *New York Times,* 30 August 1999.

67

The New York City Environmental Justice Alliance, *Our Mission* (1994).

68

See Robert Gottlieb, *Forcing the spring: the transformation of the American environmental movement* (Washington, D.C.: Island Press, 1993); Samuel P. Hays, *Beauty, health, and permanence: environmental politics in the United States, 1955–1985* (Cambridge: Cambridge University Press, 1987). By 1980 the US Congress had passed twenty major laws regulating consumer products, the environment, and workplace conditions and had established three new federal agencies, the Environmental Protection Agency (EPA), the Occupational Safety and Health Administration (OSHA), and the Council on Environmental Quality (CEQ). Yet the alliance between the labor and environmentalist movements was to be put under severe strain by economic recession in the 1980s. The US economy found itself increasingly uncompetitive within the global market economy, especially for mass-produced consumer goods, raw materials, foodstuffs, energy supplies, and capital goods, as low levels of investment combined with increasing government regulations, low productivity, low profits, and rising costs of health care, consumer product safety, and environmental protection. See Scott Dewey, "Working for the environment: organized labor and the origins of environmentalism in the United States, 1948–1970," *Environmental History* 3 (1998): 45–63; Daniel Faber and James O'Connor, "Capitalism and the crisis of environmentalism," in Richard Hofrichter, ed., *Toxic struggles: the theory and practice of environmental justice* (Philadelphia: New Society, 1993), p. 14.

69

In the case of air pollution, for example, as recently as the 1960s the city faced the emissions of eleven municipal waste incinerators along with more than seventeen thousand apartment incinerators. In subsequent decades a combination of legislative and technological change led to falling concentrations of lead, soot, and sulfur dioxide. By the 1990s the principal air pollution concerns were derived from vehicle exhaust emissions along with medical waste incinerators, sewage treatment plants, and other sources. See City of New York, *Freedom to breathe: report of the Mayor's task force;* Goldstein and Izeman, *The New York environment book.*

70

Mike Davis, *Prisoners of the American dream: politics and economy in the history of the US working class* (London: Verso, 1986), p. 179.

71

An influential report published in 1987 by the United Church of Christ provided extensive empirical evidence that race is a more significant factor than social class in the siting of hazardous waste facilities. See United Church of Christ Commission for Racial Justice, *Toxic wastes and race in the United States: a national report on the racial and socioeconomic characteristics of communities surrounding hazardous waste sites* (New York, 1987). Significant links began to emerge in the 1990s between academic research and political activism, marked in particular by the pioneering investigations of the sociologist Robert Bullard. There is now a vast literature on the US environmental justice movement. See, for example, F. O. Adeola, "Environmental hazards, health, and racial inequality in hazardous waste distribution," *Environment and Behaviour* 26 (1994): 99–126; Robert D. Bullard, ed., *Confronting environmental racism: voices from the grassroots* (Boston: South End Press, 1993); Robert D. Bullard, *Dumping in Dixie: race, class and environmental quality,* 2d ed. (Boulder, Colo.: Westview Press, 1994); Robert D. Bullard, *Unequal protection: environmental justice and communities of color* (San Francisco: Sierra Club Books, 1994); L. Burke, "Race and environmental equity: a geographic analysis in Los Angeles," *Geo Info Systems* (October 1993), 44–50; Susan L. Cutter, "Race, class and environmental justice," *Progress in Human Geography* 19 (1995): 111–122; Andrew Szasz, *Ecopopulism: toxic waste and the movement for environmental justice* (Minneapolis: University of Minnesota Press, 1994); Devon Peña, "The 'brown' and the 'green': Chicanos and environmental politics in the upper Rio Grande," *Capitalism, Nature, Socialism* 3 (1992): 79–103;

Ruth Rosen, "Who gets polluted? The movement for environmental justice," *Dissent* (Spring 1994), pp. 223–230.

72

Lois Gibbs, foreword to Hofrichter, ed., *Toxic struggles,* p. ix.

73

Faber and O'Connor, "Capitalism and the crisis of environmentalism," p. 21.

74

See, for example, Gray Brechin, "Conserving the race: natural aristocracies, eugenics, and the US conservation movement," *Antipode* 28 (1996): 229–245; Kevin Starr, *Americans and the California dream, 1850–1915* (New York: Oxford University Press, 1973). Starr, for example, describes how racial anxieties pervaded the creation of the American West rooted in the "perennial American Protestant contempt for the Latin way of life." Christopher Reed, "Green policy tainted by racial slant: a claim that immigration threatens the environment has split the ecology movement," *Guardian* (1 October 1997).

75

See Stephen Jay Gould, "An evolutionary perspective on strengths, fallacies, and confusions in the concept of native plants," in Joachim Wolschke-Bulmahn, ed., *Nature and ideology: natural garden design in the twentieth century* (Washington, D.C.: Dumbarton Oaks Research Library and Collection, 1997), pp. 11–19.

76

See Anne Whiston Spirn, "The authority of nature: conflict and confusion in landscape architecture," in Wolschke-Bulmahn, ed., *Nature and ideology,* pp. 249–261.

77

Obituary for Sydney Howe, *New York Times,* 14 April 1996.

78

William Cronon, "The trouble with wilderness; or, getting back to the wrong nature," in William Cronon, ed., *Uncommon ground: rethinking the human place in nature* (New York: W. W. Norton, 1996), p. 84.

79

Reed, "Green policy tainted by racial slant."

80

Reinerio Hernández-Marquez, Director of the Center for a Sustainable Urban Environment, Eugenio Maria de Hostos Community College, the Bronx, New York, and former Director of the North River Environmental Benefits Program, interview with the author (19 April 1996). Michelle de Pass, interview with the author (4 April 1995). Each of the organizations grouped under the New York City Environmental Justice Alliance has a unique history. The South Bronx Clean Air Coalition, for example, emerged out of conflict in the early 1990s between the Mott Haven community of the South Bronx and the developers of a medical waste incinerator, whereas the Magnolia Tree Earth Center has an earlier origin from the threat of urban renewal to Bedford-Stuyvesant. See also Dolores Greenberg, "Reconstructing race and protest: environmental justice in New York City," *Environmental History* 5 (2000): 223–250.

81

Vernice Miller, cited in Bullard, *Unequal protection.* See also Jeff Bliss, "Not just birds and trees: West Harlem environmental action," *City Limits* (May 1990), pp. 7–8; Kemba Johnson, "Green with envy," *City Limits* (January 2000), pp. 16–20; Vernice Miller, "Building on our past, planning for our future: communities of color and the quest for environmental justice," in Hofrichter, ed., *Toxic struggles,* pp. 128–136. With the recruitment of Vernice Miller to the NRDC staff in the 1990s the organization has begun to widen its involvement with environmental justice issues beyond its long-standing ecological concerns.

82

Shaw, *The activist's handbook,* p. 81.

83

"Victory claimed for areas burdened by city projects," *New York Times,* 1 May 1993. The environmental justice movement has begun to influence federal policy and in so doing has exposed the degree of the emerging tension between the American environmental movement more generally and the corporate drive for environmental deregulation in the early 1990s. Examples of the impact of the environmental justice movement include eventual establishment of a Workgroup on Environmental Equity and an Office of Environmental Equity by the Environmental Protection Agency in 1990 and signing of the Environmental Justice Executive Order 12898 by President Clinton in February 1994. See, for example, Marianne Lavelle, "Clinton pushes on race and environment," *National Law Journal* 16, no. 14 (1993); "The regulatory thickets of environmental racism," *New York Times,* 19 December 1993; US Environmental Protection Agency, *Environmental Equity: Reducing Risk for All Communities* (EPA230-R-92-008) (Washington, D.C.: Environmental Protection Agency, 1992).

84

Samme Chittum, "The politics of pollution," *City Limits* (November 1992): 18–22; New York City Department of Environmental Protection, *Greenpoint/Williamsburg environmental benefits program* (1993); Jane Sweeny, Chantel Shipman, and Anthony Tassi, *The environmental benefits program: New York City* (New York: The Mega-Cities Project/UN Development Program, 1994). The tension over industrial regeneration has become sharper with gentrification pressures spilling over from Manhattan's Lower East Side into northern Williamsburg in the late 1990s, in a process that bears similarities with the artist-led transformation of SoHo in the late 1970s. See Jonathan Bowles, "Hostile territory," *City Limits* (May 1999): 8–10. The West Harlem Environmental Benefits Program was set up after local community activists successfully sued the city for not operating the North River Sewage Treatment Plant facility within legal environmental standards. (The funding of the environmental benefits program is derived from litigation procedures between the city and the state.) The negotiations in West Harlem have been marked by community tensions, in this case between middle-class African Americans and more recent Latino immigrants from the Dominican Republic, as well as political conflict between activists seeking influence over local government policy.

85

Margrit Kaminsky, Susan Klitzman, and David Michaels, "Health profile of cancer, asthma, and childhood lead poisoning in Greenpoint/Williamsburg," First Report (New York City Department of Health/CUNY Medical School, 1992); Margrit Kaminsky, Susan Klitzman, David Michaels, and Lori Stevenson, *Health profile of cancer, birth defects, asthma, and childhood lead poisoning in Greenpoint/Williamsburg,* Second Report (New York City Department of Health/CUNY Medical School, 1993).

86

See Michael K. Heiman, "Science by the people: grassroots environmental monitoring and the debate over scientific expertise," *Journal of Planning Education and Research* 16 (1997): 291–299; Matthew Gandy, "Rethinking the ecological leviathan: environmental regulation in an age of risk," *Global Environmental Change* 9 (1999): 59–69.

87

See Kathleen McGowan, "Breathing lessons," *City Limits* (September/October 1999): 16–21, 32.

88

Carr, Zeitel, and Weiss, "Variations in asthma hospitalizations and deaths in New York City"; Richard Levins and Richard Lewontin, *The dialectical biologist* (Cambridge, Mass.: Harvard University Press, 1985).

89

Susan A. Perlin, Woodrow R. Setzer, John Creason, and Ken Sexton, "Distribution of industrial air emissions by income and race in the United States: an approach using the Toxic Release Inventory," *Environmental Science & Technology* 29 (1995): 69–80.

90

Heiman, "Science by the people."

91

Hanna Liebman, "Techno revolution," *City Limits* (February 1995), pp. 8–9; Reinerio Hernández-Marquez, interview with the author (19 April 1996); Roz Laskar, Betsy Humphreys, and William Braithwaite, *Making a powerful connection: the health of the public and the national information infrastructure* (Washington, D.C.: US Public Health Service, 1995); "Youth and duty: Williamsburg teens lend community spirit as public health advocates," *New York Daily News,* 3 August 1994; Laximi Ramasubramanian, "Building communities: GIS and participatory decision making," *Journal of Urban Technology* 3 (1995): 67–99.

92

Michael K. Heiman, personal communication with the author (16 December 1999).

93

Nancy E. Anderson, "Notes from the front line," *Fordham Urban Law Journal* 21 (1994): 757–869.

94

See, for example, Michael K. Heiman, "Race, waste, and class: new perspectives on environmental justice," *Antipode* 28 (1996): 111–121; Robert Lake, "Volunteers, NIMBYs, and environmental justice: dilemmas of democratic practice," *Antipode* 28 (1996): 160–174.

95

The Seattle riots against the World Trade Organization in December 1999 suggest that in certain circumstances the Internet can be a powerful tool for the organization of oppositional political strategies. The transformative potential of the Internet has also been advocated by Saskia Sassen and other scholars as part of the political dynamics of globalization.

96

See Paul Bradley, "Bitter justice: a South Bronx paper mill plan splinters the environmental movement," *City Limits* (October 1994), pp. 14–17. Other social and environmental impacts of recycling include the battery recycling plants of Mexico and plastics-sorting factories of South Asia. See, for example, the coverage in *Toxic Trade Update,* and V. Taliman, "Toxic waste dumping in the Third World," *Race and Class* 30 (1989): 47–56.

97

"Negatives of recycling in New York," *New York Times,* 8 August 2000.

98

"Workers pick up where New Yorkers leave off, " *New York Times,* 27 June 2000.

99

Horton, "Rethinking recycling," p. 18.

100

Charles Sabel, Archon Fung, and Bradley Karkkainen, "Beyond backyard environmentalism: how communities are quietly refashioning environmental regulation," *Boston Review* 24 (October/November 1999): 4–11.

101

Charles Sabel, Archon Fung, and Bradley Karkkainen, "Sabel, Fung, and Karkkainen respond," *Boston Review* 24 (October/November 1999): 22–23.

102

Matt Wilson and Eric Weltman, "Government's job" *Boston Review* 24 (October/November 1999): 13. Similar concerns are voiced by professor of American institutions Theodore J. Lowi, who considers Sabel et al.'s market-based localism as little more than the "decadent phase of social-democratic liberalism." Theodore J. Lowi, "Frontyard propaganda," *Boston Review* 24 (October/November 1999): 15–16.

103

Although groups such as El Puente maintain links with Third World struggles against neocolonialism and ecological genocide, the scale of grassroots political organizing has not yet matched that of other US cities such as Los Angeles. See, for example, Laura Pulido, "Restructuring and the contraction and expansion of environmental rights in the United States," *Environment and Planning A* 26 (1994): 915–936; Eric Mann, *L.A.'s lethal air: new strategies for policy, organizing, and action* (Los Angeles: A Labor/Community Strategy Center Book, 1991).

104

On the limitations of environmental justice research see Laura Pulido, "A critical review of the methodology of environmental racism research," *Antipode* 28 (1996): 142–159; Susan L. Cutter and William D. Solecki, "Setting environmental justice in space and place: acute and chronic airborne toxic releases in the southeastern United States," *Urban Geography* 17 (1996): 380–399.

105

Ira Katznelson, "Social justice, liberalism, and the city: considerations on David Harvey, John Rawls, and Karl Polanyi," in Andrew Merrifield and Erik Swyngedouw, eds., *The urbanization of injustice* (London: Lawrence and Wishart, 1998), pp. 45–64.

106

Ibid., p. 55.

107

See, for example, Susan Fainstein, "Justice, politics, and the creation of urban space," in Merrifield and Swyngedouw, eds., *The urbanization of injustice,* pp. 18–44.

108

See Ira Katznelson, *City trenches: urban politics and the patterning of class in the United States* (Chicago: University of Chicago Press, 1981), p. 198.

109

Iris Marion Young, *Justice and the politics of difference* (Princeton, N.J.: Princeton University Press, 1990), p. 234.

110

See, for example, Mike Davis, "Magical urbanism: Latinos reinvent the US big city," *New Left Review* 234 (1999): 43; David Harvey, *Justice, nature and the geography of difference* (Oxford and Cambridge, Mass.: Basil Blackwell, 1996); Florence Gardner and Simon Greer, "Crossing the river: how local struggles build a broader movement," *Antipode* 28 (1996): 175–192; Katznelson, *City trenches.*

Index

Illustrations are indicated by page numbers in italics.